DISCRETE MATHEMATICS AND GAME THEORY

THEORY AND DECISION LIBRARY

General Editors: W. Leinfellner (*Vienna*) and G. Eberlein (*Munich*)

Series A: Philosophy and Methodology of the Social Sciences

Series B: Mathematical and Statistical Methods

Series C: Game Theory, Mathematical Programming and Operations Research

Series D: System Theory, Knowledge Engineering an Problem Solving

SERIES C: GAME THEORY, MATHEMATICAL PROGRAMMING AND OPERATIONS RESEARCH

VOLUME 22

Editor: S. H. Tijs (University of Tilburg); *Editorial Board:* E.E.C. van Damme (Tilburg), H. Keiding (Copenhagen), J.-F. Mertens (Louvain-la-Neuve), H. Moulin (Durham), S. Muto (Tokyo University), T. Parthasarathy (New Delhi), B. Peleg (Jerusalem), H. Peters (Maastricht), T. E. S. Raghavan (Chicago), J. Rosenmüller (Bielefeld), A. Roth (Pittsburgh), D. Schmeidler (Tel-Aviv), R. Selten (Bonn), W. Thomson (Rochester, NY).

Scope: Particular attention is paid in this series to game theory and operations research, their formal aspects and their applications to economic, political and social sciences as well as to socio-biology. It will encourage high standards in the application of game-theoretical methods to individual and social decision making.

The titles published in this series are listed at the end of this volume.

DISCRETE MATHEMATICS AND GAME THEORY

by

Guillermo Owen

*Naval Postgraduate School,
Monterey, California, U.S.A.*

SPRINGER SCIENCE+BUSINESS MEDIA, B.V.

A C.I.P. Catalogue record for this book is available from the Library of Congress.

DOI 10.1007/978-1-4615-4991-8

Printed on acid-free paper

All Rights Reserved
© 1999 Springer Science+Business Media Dordrecht
Originally published by Kluwer Academic Publishers in 1999
MyCopy version of the original edition 1999
No part of the material protected by this copyright notice may be reproduced or utilized in any form or by any means, electronic or mechanical, including photocopying, recording or by any information storage and retrieval system, without written permission from the copyright owner.
www.springer.com/mycopy

Table of Contents

Chapter I. Vectors and Matrices — 1

1. Algebraic Operations — 1
2. Row Operations and the Solution of Systems of Linear Equations — 7
3. Solution of General m×n Systems of Equations — 15

Chapter II. Linear Programming — 27

1. Linear Programs — 27
2. The Simplex Algorithm: Slack Variables — 37
3. The Simplex Tableau — 41
4. The Simplex Algorithm: Objectives — 47
5. The Simplex Algorithm: Choice of Pivots — 48
6. The Simplex Algorithm: Stage I. — 55
7. The Simplex Algorithm: Proof of Convergence — 58
8. Equation Constraints — 66
9. Degeneracy Procedures — 70
10. Some Practical Comments — 74
11. Duality — 77
12. Transportation Problems — 92
13. Assignment Problems — 102

Chapter III. The Theory of Probability — 113

1. Probabilities — 113
2. Discrete Probability Spaces — 115
3. Conditional Probability. — 123
4. Compound Experiments — 126
5. Bayes' Formula — 131
6. Repetition of Simple Experiments; The Binomial Distribution — 138
7. Drawings with and without Replacement — 141
8. Random Variables — 145
9. Expected Values. Means and Variances — 151
10. Rules for Computing the Mean and Variance — 155
11. Two Important Theorems — 158
12. Markov Chains — 167
13. Regular and Absorbing Markov Chains — 172

Chapter IV. The Theory of Games — 181

1. Games: Extensive and Normal Form — 181

2. Saddle Points   185
3. Mixed Strategies   192
4. Solution of 2x2 Games   199
5. $2 \times n$ and $m \times 2$ Games   203
6. Solutions by Linear Programming   209
7. Solution of Games by Fictitious Play   218
8. The von Neumann Model of an Expanding Economy   222
9. Existence of an Equilibrium Expansion Rate   228
10. Two-Person Non-Zero-Sum Games   236
11. Evolutionary Stable Systems   245

Chapter V. Cooperative Games   253

1. n-Person Games   253
2. The Core   256
3. The Shapley Value   260
4. Voting Structures   266

Chapter VI. Dynamic Programming   271

1. The Principle of Maximality   271
2. The Fixed-Charge Transportation Problem   279
3. Inventories   287
4. Stochastic Inventory Systems   294

Chapter VII. Graphs and Networks   303

1. Introduction   303
2. Critical Path Analysis   304
3. The Shortest Path through a Network   314
4. Minimal Spanning Trees   319
5. The Maximal Flow in a Network   324

I. VECTORS AND MATRICES

1. ALGEBRAIC OPERATIONS

We assume that the reader is already familiar with the ordinary algebra of vectors and matrices. We give the basics below.

A matrix is a rectangular array of numbers, arranged in rows and columns. If it has m rows and n columns, we say that it is an $m \times n$ matrix. For example,

$$A = \begin{array}{|cccc|} \hline 4 & 1 & 2 & 0 \\ 2 & -4 & 5 & 1 \\ 3 & 0 & -1 & 2 \\ \hline \end{array}$$

is a 3×4 matrix, while

$$B = \begin{array}{|cc|} \hline 6 & 3 \\ 7 & 4 \\ 8 & -3 \\ -2 & 5 \\ \hline \end{array}$$

is a 4×2 matrix.

The numbers that make up the matrix are called entries. We will frequently use letters with two subscripts to represent these entries. Thus, a_{ij} would be the entry in the ith row and jth column of matrix A, b_{31} would be the entry in the 3rd row and 1st column of B, etc.

Matrices are in some sense a generalization of the concept of number. Thus it is not surprising that certain operations can be carried out on matrices, similar to the arithmetic operations. These include addition and multiplication of matrices, and scalar multiplication. There is also the operation of transposition.

Transposition. If A is an $m \times n$ matrix, then the transpose of A is an $n \times m$ matrix, B, defined by

$$b_{ij} = a_{ji}$$

We will use A' to denote the transpose of A. For example,

1.1.1
If $A = \begin{bmatrix} 5 & 7 \\ 1 & 3 \\ -4 & 2 \end{bmatrix}$ then $A' = \begin{bmatrix} 5 & 1 & -4 \\ 7 & 3 & 2 \end{bmatrix}$

Addition of matrices. If A and B are two $m \times n$ matrices (note the two matrices have the same m and the same n) then their sum, A+B, is also an $m \times n$ matrix, C, defined by

$$c_{ij} = a_{ij} + b_{ij}.$$

Thus,

$\begin{bmatrix} 3 & 4 & -1 \\ 1 & -3 & 0 \end{bmatrix} + \begin{bmatrix} 2 & -6 & 3 \\ 5 & 2 & 2 \end{bmatrix} = \begin{bmatrix} 5 & -2 & 2 \\ 6 & -1 & 2 \end{bmatrix}$

Scalar multiplication. If A is an $m \times n$ matrix, and s is a number. Then the product, sA, is also an $m \times n$ matrix, D, defined by

1.1.2 $$d_{ij} = s\, a_{ij}$$

Thus

$4 \begin{bmatrix} 3 & 4 & -1 \\ 1 & -3 & 0 \end{bmatrix} = \begin{bmatrix} 12 & 16 & -4 \\ 4 & -12 & 0 \end{bmatrix}$

Note that in this context, the number s (which multiplies the matrix) is frequently called a *scalar*.

<u>Multiplication of matrices.</u> Let A be an $m \times n$ matrix, and let B be an $n \times p$ matrix. Then the product AB is an $m \times p$ matrix, Q, defined by

1.1.3 $$q_{ik} = a_{i1}b_{1k} + a_{i2}b_{2k} + \ldots + a_{in}b_{nk}.$$

Thus,

$$\begin{bmatrix} 6 & 1 \\ 2 & 4 \\ -1 & 3 \end{bmatrix} \begin{bmatrix} 1 & 5 \\ 4 & 8 \end{bmatrix} = \begin{bmatrix} 6+4 & 30+8 \\ 2+16 & 10+32 \\ -1+12 & -5+24 \end{bmatrix} = \begin{bmatrix} 10 & 38 \\ 18 & 42 \\ 11 & 19 \end{bmatrix}$$

Note that AB is not necessarily equal to BA. In fact, it may be that one of these products exists while the other one does not; even if they both exist, moreover, they need not be equal.

I.1.1. Rules for operations on matrices

The following hold for the operations:

1. If A and B are both $m \times n$ matrices, then

$$(A + B)^t = A^t + B^t$$

2. For any A,
$$(A^t)^t = A$$

3. If A and B are both $m \times n$ matrices, then

$$A + B = B + A$$

4. If A, B, and C are three $m \times n$ matrices, then

$$(A + B) + C = A + (B + C)$$

5. If A and B are $m \times n$ matrices, and s is a scalar, then

$$s(A+B) = sA + sB$$

6. If A is a matrix and r and s are scalars, then

$$(r+s)A = rA + sA.$$

and

$$(rs)A = r(sA)$$

7. For any given *m* and *n*, there exists an *m×n* matrix, O, such that, if A is also *m×n*, then A+O = A.

8. If A is *m×n*, B is *n×p*, and C is *p×q*, then

$$(AB)C = A(BC)$$

9. If A is *m×n*, B is *n×p*, and *s* is a scalar, then

$$(sA)B = A(sB) = s(AB)$$

10. If A and B are both *m×n* n, and C is *n×p*, then

$$(A + B) C = AC + BC$$

11. If A is *m×n*, and B and C are both *n×p*, then

$$A (B + C) = AB + AC$$

12. For any *n*, there exists and *n×n* matrix, I, such that, if A is any *m×n* matrix, then

$$AI = A$$

and, if B is any *n×p* matrix, then

$$IB = B.$$

13. If A is *m×n*, and B is *n×p*, then

$$(AB)^t = B^t A^t$$

The matrix O, referred to in (7) above, is simply an *m×n* matrix, all of whose entries are 0. It is known as the additive identity, or zero matrix. Note that for any A, we can define the matrix -A as the product (-1)A. It is easy to check that

$$A + (-A) = -A + A = O$$

and we say that -A is the additive inverse, or negative, of A.

The matrix I, referred to in (12) above, is an $n \times n$ matrix with entries δ_{ij}, where $\delta_{ij} = 1$ if $i = j$, and 0 if $i \neq j$, for example (in case $n = 3$):

$$I = \begin{bmatrix} 1 & 0 & 0 \\ 0 & 1 & 0 \\ 0 & 0 & 1 \end{bmatrix}$$

This matrix is known as the unit matrix or multiplicative identity (or order n).

If A is an $n \times n$ matrix, it may be that there exists another $n \times n$ matrix, B, such that

$$AB = BA = I.$$

If so, we say that B is the multiplicative inverse (more simply, inverse) of A, and we denote it by A^{-1}. We also say that A is invertible, or non-singular. Note that, if A is not square (i.e., $m \neq n$), it does not have an inverse, and it cannot be non-singular. But even a square matrix need not have an inverse.

As an example,

$$\text{if } A = \begin{bmatrix} 5 & 2 \\ 3 & 1 \end{bmatrix} \text{ and } B = \begin{bmatrix} 1 & -2 \\ -3 & 5 \end{bmatrix}$$

it is easy to check that $AB = BA = I$, so that A and B are inverses of each other.

Vectors

The easiest way to define a vector is to say that it is a matrix with either one row or one column. Then, a $1 \times n$ matrix is a row-vector, while an $m \times 1$ matrix is a column vector. We shall deal mostly with column vectors, and it is understood that a vector is a column vector unless we specify otherwise. The entries in a vector are called components.

This definition says that a vector is a special type of matrix. However, from our point of view, vectors and matrices represent fundamentally different types of entities. In essence, a vector is, for us, a point in m-dimensional space (where m can be any positive integer). We will represent this space by $\Re^m$.

For $m = 2$ or $m = 3$, it is easy to understand what we mean by m-dimensional space. For larger values of m, we are of course not speaking about the usual space of physics; rather, we can think of an m-vector as representing the values of m quantities. The meaning will depend on context. Thus, if we are talking about apples, pears, and bananas, the vector $\mathbf{v} = (3,5,1)^t$ might mean three apples, 5 pears, and 1 banana. (Of course, if we were talking about prices at the supermarket, it might also mean that apples, pears and bananas were selling for \$3, \$5, and \$1 per kilo, respectively.)

A matrix, on the other hand, represents for us what we call a *linear transformation*. In effect, note that if $\mathbf{v}$ is an n-vector, and A is an $m \times n$ matrix, then the product $A\mathbf{v}$ is an m-vector. Thus, we say that the matrix A (or, more precisely, multiplication by A) gives us a transformation from $\Re^n$ into $\Re^m$: it transforms the n-vector $\mathbf{v}$ into the m-vector $A\mathbf{v}$. By the rules given above, we find that if A is a matrix, $\mathbf{v}$ and $\mathbf{w}$ are vectors, and s is a scalar, then (assuming these operations can be carried out),

1.1.4 $$A(\mathbf{v}+\mathbf{w}) = A\mathbf{v} + A\mathbf{w}$$

1.1.5 $$A(s\mathbf{v}) = s(A\mathbf{v})$$

Transformations which satisfy these two conditions are called linear transformations.

As an example, consider the matrix

$$A = \begin{bmatrix} 1 & -2 \\ 6 & 5 \\ -3 & 4 \end{bmatrix}$$

Let $\mathbf{v}$ be the (column) 2-vector with components x and y. It is easy to see that A transforms $\mathbf{v}$ into the 3-vector with components

$$r = x - 2y$$
$$s = 6x + 5y$$
$$t = -3x + 4y$$

2. ROW OPERATIONS AND THE SOLUTION OF SYSTEMS OF LINEAR EQUATIONS

Let us consider a system of two linear equations in three unknowns:

$$2x + 5y - 4z = 6$$
$$x - 2y + 4z = 3$$

It may be seen that this can be rewritten in matrix notation as

$$\begin{bmatrix} 2 & 5 & -4 \\ 1 & -2 & 2 \end{bmatrix} \begin{bmatrix} x \\ y \\ z \end{bmatrix} = \begin{bmatrix} 6 \\ 3 \end{bmatrix}$$

More generally, the system of m linear equations in n unknowns,

1.2.1
$$a_{11}x_1 + a_{12}x_2 + \ldots + a_{1n}x_n = c_1$$
$$a_{21}x_1 + a_{22}x_2 + \ldots + a_{2n}x_n = c_2$$
$$\ldots \ldots \ldots \ldots \ldots \ldots \ldots \ldots$$
$$a_{m1}x_1 + a_{m2}x_2 + \ldots + a_{mn}x_n = c_m$$

can be rewritten in the form

1.2.2
$$\begin{bmatrix} a_{11} & a_{12} & \ldots & a_{1n} \\ a_{21} & a_{22} & \ldots & a_{2n} \\ \ldots & \ldots & \ldots & \ldots \\ \ldots & \ldots & \ldots & \ldots \\ \ldots & \ldots & \ldots & \ldots \\ a_{m1} & a_{m2} & \ldots & a_{mn} \end{bmatrix} \begin{bmatrix} x_1 \\ x_2 \\ \ldots \\ \ldots \\ \ldots \\ x_n \end{bmatrix} = \begin{bmatrix} c_1 \\ c_2 \\ \ldots \\ \ldots \\ \ldots \\ c_m \end{bmatrix}$$

or, more simply, as

1.2.3
$$Ax = c$$

where A is the matrix of coefficients, **x** is the column vector of variables, and C is the vector of constant terms. Let K be any $m \times m$ matrix. If we multiply both sides of (1.2.3) by K, we obtain, by the associative law,

1.2.4 $$(KA)\mathbf{x} = K\mathbf{c}$$

and it is clear that (1.2.3) implies (1.2.4), i.e., any vector **x** that satisfies (1.2.3) will also satisfy (1.2.4). The converse is not, generally, true: the solutions of (1.2.4) need not be solutions to (1.2.3). If, however, K is invertible, we may multiply both sides of (1.2.4) by K^{-1}:

$$K^{-1}(KA)\mathbf{x} = K^{-1}K\mathbf{c}$$

which reduces to (1.2.3). Thus, in this case, (1.2.4) implies (1.2.3), and so the two matrix equations, (1.2.3) and (1.2.4), are equivalent: any vector that solves the one will also solve the other. Now, it must be admitted that (1.2.4) has approximately the same form as (1.2.3): for an arbitrary, non-singular matrix K, we have no reason to believe that the solution of the matrix equation (1.2.4) will be any easier than that of (1.2.3). We shall see, however, that the matrix K can be chosen in such a way that the equation (1.2.4) is either trivially easy to solve or else shows clearly that no solution is possible. More exactly, we shall show
the existence of sequence, $K_1, K_2, \ldots K_p$, of non-singular matrices such that the equation

$$K_p \ldots K_2 K_1 A \mathbf{x} = K_p \ldots K_2 K_1 \mathbf{c}$$

can be easily solved if it has any solution at all.

Let us consider the following operations that may be performed on any matrix:

1.2.5 Multiply every entry in the kth row by the non-zero scalar, r; i.e., replace a_{ks} by ra_{ks}.

1.2.6 Multiply each entry in the lth row by r and add to the corresponding entry in the kth row, i.e., replace a_{kj} by $a_{kj} + ra_{lj}$ (where $k \neq l$).

These two operations are called *fundamental row* operations; examples of them are:

1.2.1. Examples

$$\begin{bmatrix} 1 & 5 & 2 \\ -1 & 4 & 6 \\ 2 & 1 & 8 \end{bmatrix} \rightarrow \begin{bmatrix} 1 & 5 & 2 \\ 3(-1) & 3(4) & 3(6) \\ 2 & 1 & 8 \end{bmatrix} = \begin{bmatrix} 1 & 5 & 2 \\ -3 & 12 & 18 \\ 2 & 1 & 8 \end{bmatrix}$$

In this example, the second row was multiplied by the scalar 3.

$$\begin{bmatrix} 1 & -1 & 6 \\ 2 & 4 & 5 \\ 3 & 1 & 8 \end{bmatrix} \rightarrow \begin{bmatrix} 1+4 & -1+8 & 6+10 \\ 2 & 4 & 5 \\ 3 & 1 & 8 \end{bmatrix} = \begin{bmatrix} 3 & 7 & 16 \\ 2 & 4 & 5 \\ 3 & 1 & 8 \end{bmatrix}$$

Here, the second row was multiplied by 2, and added to the first row.

$$\begin{bmatrix} 1 & 4 & 7 \\ 1 & 4 & 6 \\ 4 & 3 & 2 \end{bmatrix} \rightarrow \begin{bmatrix} 1 & 4 & 7 \\ 1-1 & 4-4 & 6-7 \\ 4 & 3 & 2 \end{bmatrix} = \begin{bmatrix} 1 & 4 & 7 \\ 0 & 0 & -1 \\ 4 & 3 & 2 \end{bmatrix}$$

In this example, the first row was multiplied by -1, and added to the second row, i.e., the first row was subtracted from the second.

It is our desire to use these row operations to solve systems of linear equations. To do this, we shall show, first, that these operations are always equivalent to multiplication on the left by a non-singular matrix K, and second, that the matrix K depends only on the form of the operation, and not on the particular matrix A upon which we operate. This second property means that multiplication by K will perform the same operation on the vector **c** as on the matrix A.

Consider, then, operation (2.6.5). It may be seen that the corresponding matrix is T = (t_{ij}), defined by

1.2.7
$$t_{ij} = \begin{cases} r & \text{if } i = j = k \\ 1 & \text{if } i = j \neq k \\ 0 & \text{if } i \neq j. \end{cases}$$

(T is almost an identity matrix, the only difference being that its (k,k)th entry is r instead of 1.) The matrix is non-singular; its inverse is $T^{-1} = (t'_{ij})$, where

1.2.8
$$t'_{ij} = \begin{cases} 1/r & \text{if } i = j = k \\ 1 & \text{if } i = j \neq k \\ 0 & \text{if } i \neq j. \end{cases}$$

That T will, in fact, cause the operation (1.2.5) can be easily proved and is left as an exercise to the reader. That (1.2.8) does, in **fact,** define the inverse matrix is most easily seen when we consider that it induces the opposite operation, i.e., it causes the kth row to be *divided* by the scalar r.

Next let us consider operation (1.2.6). It may be seen that this is induced by the matrix I + S, where I is the identity matrix, and S = (s_{ij}) is defined by

1.2.9
$$s_{ij} = \begin{cases} r & \text{if } i = k \text{ and } j = l \\ 0 & \text{otherwise} \end{cases}$$

(S has all entries equal to zero except for the (k,l)th entry, which is equal to r.) It is not difficult to see, since $k \neq l$, that SS = O, and so

$$\begin{aligned}(I + S)(I - S) &= II + SI - IS + SS \\ &= I + S - S + O \\ &= I\end{aligned}$$

Similarly, we can see that (I-S)(I+S) = I, so that I-S is the inverse of I+S, and so I+S is non-singular.

We are now in a position to apply these elementary row operations to the solution of systems of linear equations. The general idea is to perform the operations on the matrix of coefficients until a very simple type of matrix is obtained; we try to change A, if possible, into an identity matrix. If A is singular, this will not be feasible, so we have to content ourselves with something that looks as much like in identity matrix as possible.

We have seen that each of these row operations is equivalent to multiplication by a non-singular matrix K. The vector of constant terms, **c**, must simultaneously be multiplied by the same K. But K induces the same row operation on all matrices (so long as they have the proper number of rows so that multiplication is possible). Hence we need only perform the same row operations on the vector **c** as on the matrix A.

Since the same row operations are to be performed on both A and C, it is generally more convenient to deal with the *augmented* matrix

1.2.10

$$(A|\mathbf{c}) = \begin{array}{|ccccc|c|} \hline a_{11} & a_{12} & \cdots & a_{1n} & & c_1 \\ a_{21} & a_{22} & \cdots & a_{2n} & & c_2 \\ \cdots & \cdots & \cdots & \cdots & & \cdots \\ \cdots & \cdots & \cdots & \cdots & & \cdots \\ \cdots & \cdots & \cdots & \cdots & & \cdots \\ a_{m1} & a_{m2} & \cdots & a_{mn} & & c_m \\ \hline \end{array}$$

It is then simply a question of performing the row operations on $(A|\mathbf{c})$; this guarantees that they be performed simultaneously on A and **c**.

1.2.2. Example. Solve the system of equations

$$\begin{aligned} x + 2y - z + w &= 5 \\ 2x - y \quad\quad + w &= 6 \\ x - y + z + 3w &= 9 \\ 4y - z - 2w &= 1 \end{aligned}$$

This system may be written, using matrix notation, in the form

$$\begin{bmatrix} 1 & 2 & -1 & 1 \\ 2 & -1 & 0 & 1 \\ 1 & -1 & 1 & 3 \\ 0 & 4 & -1 & -2 \end{bmatrix} \begin{bmatrix} x \\ y \\ z \\ w \end{bmatrix} = \begin{bmatrix} 5 \\ 6 \\ 9 \\ 1 \end{bmatrix}$$

We obtain, thus, the augmented matrix

$$\begin{bmatrix} 1 & 2 & -1 & 1 & | & 5 \\ 2 & -1 & 0 & 1 & | & 6 \\ 1 & -1 & 1 & 3 & | & 9 \\ 0 & 4 & -1 & -2 & | & 1 \end{bmatrix}$$

We would like, if possible, to change this matrix, by means of row operations, until the first four columns form an identity matrix. The best method is, generally, to take the columns one at a time, changing the "main diagonal" entry to a 1 and getting rid of any other non-zero entries in the column. In the first column, here, we find that the first entry is, as desired, a 1. The entries in the second and third rows are not zero, so we must eliminate them. To eliminate the 2 in the second row, we can multiply the first row by -2 and add to the second row; we also subtract the first row from the third, obtaining

1	2	-1	1	5
2-2	-1-4	0+2	1-2	6-10
1-1	-1-2	1+1	3-1	9-5
0	4	-1	-2	1

=

1	2	-1	1	5
0	-5	2	-1	-4
0	-3	2	2	4
0	4	-1	-2	1

We have, thus, reduced the first column to the required form. We now attack the second column: we must obtain here a 1 in the second row and zeros elsewhere, *being careful, meanwhile, that the first column remains as it is.* (This means that we cannot, say, multiply the first row by 3 and add to the second row; while this would accomplish the feat of putting a 1 in the second row, second column, it would. put a 3 in the second row, first column.)

There are many ways of putting a 1 in the second row, second column. One possibility is to multiply the third row by 2 and add to the second row. Another possibility is to multiply the second row by -1/5. The former procedure has the advantage of avoiding fractions, at least for the moment. The latter has the advantage of being somewhat more systematic. Let us, then, carry out this latter procedure; we obtain the matrix

1	2	-1	1	5
0	1	-2/5	1/5	4/5
0	-3	2	2	4
0	4	-1	-2	1

We now use the second row to remove all the other non-zero entries in this column. We can multiply the second row by -2, and add to the first row; by 3, and add to the third row; by -4, and add to the fourth row. Carrying out all these operations, we will have

1	0	-1/5	3/5	17/5
0	1	-2/5	1/5	4/5
0	0	4/5	13/5	32/5
0	0	3/5	-14/5	-11/5

We proceed to the third column. We can multiply the third row by 4/5:

1	0	-1/5	3/5	17/5
0	1	-2/5	1/5	4/5
0	0	1	13/4	8
0	0	3/5	-14/5	-11/5

and now we use the third row to remove the other non-zero entries in the third column, obtaining

1	0	0	5/4	5
0	1	0	3/2	4
0	0	1	13/4	8
0	0	0	-19/5	-7

We next multiply the bottom row by -4/19:

1	0	0	5/4	5
0	1	0	3/2	4
0	0	1	13/4	8
0	0	0	1	28/19

and use this row to eliminate the other fourth-column entries:

1	0	0	0	60/19
0	1	0	0	34/19
0	0	1	0	61/19
0	0	0	1	28/19

We have now accomplished what we set out to do: the matrix of coefficients has become an identity matrix, and the original matrix equation, $A\mathbf{x} = \mathbf{c}$, has been reduced to the form

$$I \begin{bmatrix} x \\ y \\ z \\ w \end{bmatrix} = \begin{bmatrix} 60/19 \\ 34/19 \\ 61/19 \\ 28/19 \end{bmatrix}$$

But the matrix I will not change anything. This last equation gives us, directly, the solution of the original system, $(x, y, z, w) = (60/19, 34/19, 61/19, 28/19)$. That this satisfies the original system may be checked directly.

II.6.3. Example. Solve the system

$$\begin{aligned} x - 2y + z + 2w &= 18 \\ 2x + y + 4z - w &= 16 \\ x + y - z + w &= 4 \\ x + 3y - 2z + w &= 1 \end{aligned}$$

As before, we form the augmented matrix

1	-2	1	2	18
2	1	4	-1	16
1	1	-1	1	4
1	3	-2	1	1

We use the first row to remove the other non-zero entries in the first column:

1	-2	1	2	18
0	5	2	-5	-20
0	3	-2	-1	-14
0	5	-3	-1	-17

Next, we divide the second row by 5 and use this row to remove the non-zero entries in the second column:

1	0	9/5	0	10
0	1	2/5	-1	-4
0	0	-16/5	2	-2
0	0	-5	4	3

We multiply the third row by -5/16 and use it to clear the third column:

1	0	0	9/8	71/8
0	1	0	-3/4	-17/4
0	0	1	-5/8	5/8
0	0	0	7/8	49/8

Finally, we multiply the last row by 8/7 and clear the fourth column,

1	0	0	0	1
0	1	0	0	1
0	0	1	0	5
0	0	0	1	7

This gives us the solution; it is *(x, y, z, w)* = (1, 1, 5, 7). Again, we can check directly by substituting these values in the original equations.

3. SOLUTION OF GENERAL $M \times N$ SYSTEMS OF EQUATIONS

Naturally, it is not always possible to solve systems of equations as we did in Examples I.2.2 and I.2.3. Even if the matrix of coefficients is square, we have no guarantee that it will be invertible. We can see what happens in such cases from the following examples.

I.3.1. Example. Solve the system

$$x + 2y - z = 6$$
$$3x - y + z = 8$$
$$5x + 3y - z = 20$$

In this case, we form the augmented matrix

1	2	-1	6
3	-1	1	8
5	3	-1	20

and use the first row to clear the first column:

1	2	-1	6
0	-7	4	0
0	-7	4	0

It may be seen that the second and third rows have become identical; this means that one of the equations is redundant, so that, with only two independent equations, we cannot hope to have a unique solution. If we continue with our procedure, multiplying the second row by -1/7 and clearing the second column, we obtain

1	0	1/7	22/7
0	1	-4/7	10/7
0	0	0	0

Consider this new matrix: we would like to get a 1 in the third row, third column. This cannot be done by multiplying the third row, since we now have a zero in that position. We could do this by using one of the other rows, say, by multiplying the first row by 7 and adding to the third row. If we do this, however, we obtain a 7 in the third row, first column, thus undoing some of our previous work. We find, in fact, that we can go no further in simplifying the coefficient matrix. Our final result is the system

$$x + 1/7\, z = 22/7$$
$$y - 4/7\, z = 10/7$$

The third equation, of course, reduces to the identity $0 = 0$, and can be dispensed with. We can, then, solve for x and y in terms of z:

$$x = 22/7 - 1/7\, z$$
$$y = 10/7 + 4/7\, z$$

The variable z is arbitrary. This is the general solution, or, more exactly, one form of the general solution. We could, of course, solve for any two of the variables in terms of the third. Let us suppose that we want to solve for y and z in terms of x. We can do this by obtaining a second-order identity matrix in the columns corresponding to y and z, rather than in those corresponding to x and y. Starting from the last of the foregoing matrices, we can multiply the first row by 7:

7	0	1	22
0	1	-4/7	10/7
0	0	0	0

We then clear the third column by means of the first row:

7	0	1	22
4	1	0	14
0	0	0	0

We could, of course, have discarded the third row a long time ago, reducing to the 2 x 4 matrix

7	0	1	22
4	1	0	14

Now, the two columns corresponding to y and z do not form an identity matrix, but this is not too important. In fact, these two columns form what is called a *permutation* matrix, a matrix that can be transformed into an identity matrix by simply interchanging some of the rows. As such, the foregoing matrix is sufficient for our purposes. Indeed, the equations we now have

$$x + z = 22$$
$$4x + y = 14$$

or, solving for y and z in terms of x,

$$y = 14 - 4x$$
$$z = 22 - 7x$$

which, for arbitrary x, is another form of the general solution to the system.

I.3.2.Example. Solve the system

$$x + 2y - z = 6$$
$$3x - y + z = 8$$
$$5x + 3y - z = 22$$

Again, we form the augmented matrix

1	2	-1	6
3	-1	1	8
5	3	-1	22

and use the first row to clear the first column:

1	2	-1	6
0	-7	4	-10
0	-7	4	-8

We then divide the second row by -7, and clear the second column, obtaining

1	0	1/7	22/7
0	1	-4/7	10/7
0	0	0	2

Consider, now, the third row: it represents the contradiction $0 = 2$. This means that the system has no solution. In general, a system will be infeasible (i.e., have no solution) if a row is obtained in which all the entries except the last are zero. On the other hand, as we saw from Example I.3.1, a row that has all its entries (including the last one) equal to zero may be discarded and does not affect the system.

For systems that have fewer equations than variables, the procedure will generally be as in Example I.3.1: we try to obtain an identity matrix (or a permutation matrix) in the columns corresponding to some of the variables; this solves the system for the corresponding variables in terms of the others.

I.3.3. Example. Solve the system

$$2x + y - 4z = 10$$
$$x - y + 2z = 6$$

We can, here, attempt to solve the system for any two of the variables in terms of the third. (In this case, our attempts will be successful, but in some cases they need not be.) Let us, then, solve for x and z in terms of y. We will work on the matrix in such a way that the columns corresponding to x and z form an identity (or permutation) matrix. The augmented matrix is

2	1	-4	10
1	-1	2	6

We can divide the first row by 2 and clear the first column:

1	1/2	-2	5
0	-3/2	4	1

Next we divide the second row by 4 and clear the third column to obtain

1	-1/4	0	11/2
0	-3/8	1	1/4

so that the desired solution is

$$x = 11/2 + 1/4\ y$$
$$z = 1/4 + 3/8\ y$$

Let us suppose, now, that we want to solve for y and z in terms of x. In this last matrix, we multiply the first row by -4:

-4	1	0	-22
0	-3/8	1	1/4

and use the first row to clear the second column:

-4	1	0	-22
-3/2	0	1	-8

thus obtaining the solution

$$y = -22 + 4x$$
$$z = -8 + 3/2\, x$$

11.7.4I.3.4. Example. Solve the system

$$2x + y + 4z = 10$$
$$x - y + 2z = 6$$

Suppose we want to solve this system for x and z in terms of y. We take the augmented matrix

2	1	4	10
1	-1	2	6

and, as before, divide the first row by 2, and clear the first column:

1	1/2	2	5
0	-3/2	0	1

We would like to clear the third column now, but it has become impossible. In fact, we cannot solve for x and z in terms of y. The reason for this is best seen if we note that the second row represents the equation

$$-3/2\, y = 1$$

which tells us that $y = -2/3$. In effect, we have found that y is not arbitrary; if we were to give x and z in terms of y, we would get only one solution. But the problem has an infinity of solutions.

While it is not possible, here, to solve for x and z in terms of y, we may, if we wish, solve for x and y in terms of z. **In the last matrix** given, we may multiply the second row by $-2/3$, and clear the second column:

1	0	-2	16/3
0	1	0	-2/3

which gives us

$$x = 16/3 + 2z$$
$$y = -2/3$$

Note that x is given in terms of z; y, however, has only one value.

In case we have more equations than variables, the usual procedure is to try to obtain an identity matrix in the first n rows and n columns; below this, we try to get nothing but zeros. When this has been done, we might find that the last $m-n$ rows contain zeros only; if so, we have obtained the solution. On the other hand, it may be that one of the last $m-n$ rows has a non-zero entry in the right-hand column, but all zeros otherwise. In this case, the problem has no solution, as this row represents the contradiction $0 = 1$.

1.3.5. Example. Solve the system

$$x + y + z = 9$$
$$2x - y + 2z = 15$$
$$x - y + 2z = 12$$
$$x + 2y - z = 0$$

We take the augmented matrix

1	1	1	9
2	-1	2	15
1	-1	2	12
1	2	-1	0

and use the first row to clear the first column:

1	1	1	9
0	-3	0	-3
0	-2	1	3
0	1	-2	-9

We divide the second row by -3, and use it to clear the second column:

1	0	1	8
0	1	0	1
0	0	1	5
0	0	-2	-10

Finally, we can use the third row to clear the second column, obtaining the matrix

1	0	0	3
0	1	0	1
0	0	1	5
0	0	0	0

which gives us the solution

$$x = 3$$
$$y = 1$$
$$z = 5$$

Note that the last row consists only of zeros; it may be disregarded.

I.3.6. Example. Solve the system

$$x + y - z = 1$$
$$2x + y + z = 9$$
$$x - 2y + z = 1$$
$$x + 2y - 2z = 0$$
$$x - y + z = 2$$

Let us once again take the augmented matrix

1	1	-1	1
2	1	1	9
1	-2	1	1
1	2	-2	0
1	-1	1	2

Once again, we use the first row to clear the first column:

1	1	-1	1
0	-1	3	7
0	-3	2	0
0	1	-1	-1
0	-2	2	1

We multiply the second row by -1, and clear the second column:

1	0	2	8
0	1	-3	-7
0	0	-7	-21
0	0	2	6
0	0	-4	-13

Finally, we divide the third row by -7, and clear the third column:

1	0	0	2
0	1	0	2
0	0	1	3
0	0	0	0
0	0	0	-1

Consider this matrix: the first three rows seem to give us the solution $(x, y, z) = (2, 2, 3)$. The last row, however, represents the contradiction $0 = 1$. We conclude that the system is infeasible, since the unique solution to the first three equations fails to satisfy the fifth equation.

PROBLEMS ON SOLUTION OF GENERAL $m \times n$ SYSTEMS OF EQUATIONS

1. Solve the following systems of simultaneous equations:

(a) $3x + 5y = 6$
$x - 2y = 2$

(b) $3x + 5y = 4$
$x + 2y = 7$

(c) $5x + 2y + 4z = 7$
$3x - 2y + z = 6$
$x + y + z = 4$

(d) $2x - 6y + z = -11$
$x + 2y - z = 2$
$-x + y + 2z = 12$

(e) $x - 2y + 3z - w = 12$
$2x + y + z - 2w = -6$
$-x + y + 2z + w = 2$
$-y + z + w = 1$

(f) $3x + y + z = 2$
$x - 2y + 2z = 5$
$2x - y - z = 7$
$5x + 3y + z = 2$

(g) $x + 2y + 4z + w = 6$
$3x + 2y - 3z - w = 5$
$-x + 6y + 2z + 3w = 8$

(h) $2x + 3y + z - 2w = 0$
$x - 2y + 2z - w = 9$
$3x - y + 3z + w = 6$
$-x + y - z + 3w = 2$

(i) $3x + 2y + 6z + w = 12$
$x - 2y + 3z + 2w = 4$
$-2x + 4y - 2z - w = -1$
$x + y - 2z + 3w = 3$

(j) $2x + 4y + 2z - w = 18$
$x + 3y - 2z + w = 13$
$-x + 2y + 2w = 28$
$3x + y + 4z = 1$

(k) $x + 2y + 6z - w = 5$
$3x + y + 4z - 2w = 7$
$x - y + 7z + w = 9$
$-x + 2y + 3z - 2w = 5$

(l) $x - 3y + 5z - w = -3$
$2x + y + 4z + w = 11$
$-x + y + z - 3w = -7$
$3x + 6y + 2z - 4w = 5$
$2x + 4y + z - 7w = -10$
$x + y + z - w = 1$

(m) $\quad x - 2y + 3z + w = 6$
$\quad\quad 2x + y + 6z - 2w = 1$
$\quad\quad 5x + 2y - 3z - 2w = 4$
$\quad\quad 4x + y - 2z - w = -1$
$\quad\quad 6x + 2y + z + w = 10$

(n) $\quad 2x + 3y + z - 5w = 8$
$\quad\quad -x + 3y + 7z + 5w = 30$
$\quad\quad x - 2y + 4z - 6w = 7$
$\quad\quad -2x + y + z + 8w = 11$

(o) $\quad 3x + y + 2z - w = 6$
$\quad\quad -x - y + 2z - 2w = 5$
$\quad\quad x + 4y + 6z - 3w = 8$

II. LINEAR PROGRAMMING

1. LINEAR PROGRAMS

In Chapter 1, we studied systems of linear inequalities and saw that in general, such systems do not have unique solutions. Nor is it generally possible to characterize their solutions as simply as those of systems of linear equations. This is not, however, an important consideration, for it is seldom desired to find *all* the solutions to such a system. It is usually only necessary to find *one* solution, i.e., a point that satisfies the constraints (inequalities) of the system. In a few cases, any solution will do, but normally, a special solution is required-the one that is best in a particular sense, say, that maximizes profits or minimizes costs. We can see this best from the following example.

II.1.1. Example. A dietitian must prepare a mixture of foods A, B, and C. Each unit of food A contains 3 oz. protein, 2 oz. carbohydrate, and 6 oz. fat, and costs 60¢. Each unit of food B contains 5 oz. protein, 6 oz. carbohydrate, and 2 oz. fat, and costs $1.25. Each unit of food C contains 3 oz. protein, 4 oz. carbohydrate, and 5 oz. fat and costs 50¢. The mixture must contain at least 20 oz. protein, 18 oz. carbohydrate, and 30 oz. fat. It is desired to find the mixture of minimal total cost, subject to these constraints.

Letting x, y, and z be the amounts of foods A, B, and C, respectively, and letting w represent the total cost of the mixture, we can summarize the data of the problem in a table (Table III.1.1).

TABLE III.1.1

Food	Amount	Protein	Carbohydrate	Fat	Cost ($)
A	x	3	2	6	0.60
B	y	5	6	2	1.25
C	z	3	4	5	0.50
Total		≥ 20	≥ 18	≥ 30	

We may express the problem in the following form:

Minimize
$$w = 0.60x + 1.25y + 0.50z$$

Subject to
$$3x+5y+3z \geq 20$$
$$2x+6y+4z \geq 18$$
$$6x+2y+5z \geq 30$$
$$x, y, z \geq 0$$

A problem such as this, of maximizing or minimizing a linear function of the variables, subject to linear constraints (equations or, more commonly, inequalities), is called a *linear program*. It is not possible, at this point, to solve the example since we have not yet developed the necessary mathematical tools. We will, therefore, give a strict mathematical analysis of the subject and come back to the example later on.

The science of linear programming is a relatively new branch of mathematics; it was among the many studies that were given impetus by the interest of military planners during World War II. (Prior to this war, only trial-and-error and approximation techniques were available for the solution of linear programs.) The most commonly used method nowadays, the *simplex algorithm,* was developed by G. B. Dantzig during the early 1950's; certain refinements in technique and notation as well as the theory of duality were later introduced by A. W. Tucker.

The standard form of the linear programming problem may be given as

2.1.1 Maximize

$$w = c_1x_1 + c_2x_2 + \ldots + c_n x_n$$

Subject to

2.1.2
$$a_{11}x_1 + a_{12}x_2 + \ldots + a_{1n}x_n \leq b_1$$
$$a_{21}x_1 + a_{22}x_2 + \ldots + a_{2n}x_n \leq b_2$$
$$\ldots$$
$$\ldots$$
$$a_{m1}x_1 + a_{m2}x_2 + \ldots + a_{mn}x_n \leq b_m$$

2.1.3
$$x_1, x_2, \ldots, x_n \geq 0$$

or, in matrix notation:

> Maximize

2.1.4 $$w = c^t x$$

> Subject to

2.1.5 $$Ax \leq b$$

2.1.6 $$x \geq 0$$

where $A = (a_{ij})$ is the matrix of coefficients, **b** and **c** are constant column vectors, **x** is the column vector of variables, and **0** is the zero column vector. (We add that vector inequalities hold whenever the inequalities hold, component by component)

We give some definitions. The variable w, given by (2.1.1) or (2.1.4), is called the *objective function*. The inequalities (2.1.2) and (2.1.3) or (2.1.5) and (2.1.6) are the *constraints* of the problem. The set of points satisfying the constraints is called the *constraint set*. Such points are *feasible* points.

Let us suppose that there is a vector, **x***, that satisfies the constraints and such that, if **x** is any vector in the constraint set, $c^t x \leq c^t x^*$. In this case we say that **x*** is the *solution* of the program (2.1.1) through (2.1.3) or (2.1.4) through (2.1.6); while $c^t x^*$ is the *value* of the program.

Strictly speaking, the program (2.1.1) through (2.1.3) is not the most general type of linear program. It may be, for example, that the objective function is to be minimized rather than maximized; that some of the constraints have opposite sign, $\geq$ instead of $\leq$; or even that some of the constraints may be equations instead of inequalities. It may also happen that some of the variables are allowed to have negative values. We can always, however, reduce a program to the standard form of (2.1.1) through (2.1.3). In fact, a function can be minimized by maximizing its negative. If an inequality has the form $\geq$, we can reverse it through the simple expedient of multiplying by –1. In turn, an equation $\alpha = \beta$, may be replaced by the two inequalities $\alpha \leq \beta$ and $-\alpha \leq -\beta$. Finally, an unrestricted variable (i.e., one that is allowed to have negative values) may be expressed as the difference of two non-negative variables. The program is, hence, as general as we desire; we are concerned, mainly, with programs that have this form, though occasionally problems of other forms may be considered without reduction to the standard form.

Let us analyze the problem by looking at the geometry of the situation. The first thing to be pointed out is that the constraint set is *convex*. Geometrically, a set, S, is convex if, whenever the points P and Q both lie in S, then the entire line-segment determined by these two points (i.e., every point that lies on the line PQ, between P and Q, lies in the set S). Figures II.1.1 (a) and (b) show examples of convex sets. On the other hand Figure II.1.1(c) shows an example of a set that is not convex.

FIGURE II.1.1 (a) and (b) are convex; (c) is not.

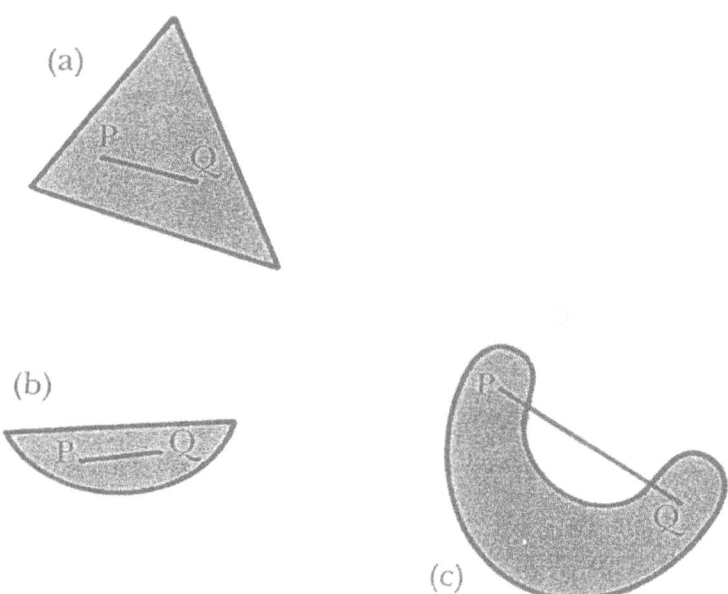

To prove that the constraint set for a linear program is convex, we recall that each of the constraints (2.1.2) or (2.1.3) is satisfied by the set of points lying on one side of a hyperplane. It is clear that such a set is convex, since, if two points lie above a plane, then so does any point between them. Suppose that P and Q both satisfy all the constraints, and R is any point between them (Figure II.1.2). It is clear that R will lie on the "correct" side of each of the hyperplanes, since P and Q do. But this means that R lies in the constraint set Thus, the constraint set is convex.

The second thing to be pointed out is that the objective function is linear:

$$w = c_1 x_1 + c_2 x_2 + \ldots + c_n x_n$$

Now, consider any line in n-dimensional space. As we know from previous study, such a line can be expressed by making one of the variables, say x_1, arbitrary and expressing all the other variables in terms of it:

2.1.7
$$x_j = \alpha_j x_1 + \beta_j$$

FIGURE II.1.2 The set of points that satisfy a collection of linear inequalities is convex.

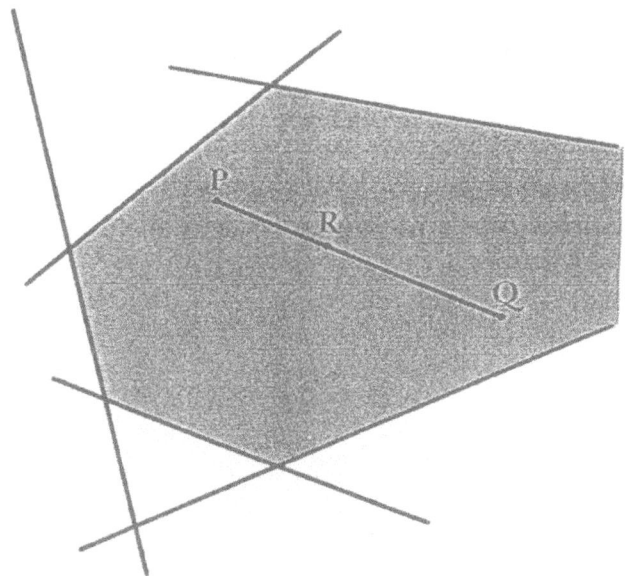

where the α_j and β_j are constants, for each value of $j = 2, 3, ..., n$. If we substitute (2.1.7) into (2.1.1), we obtain

$$w = (c_i + \sum c_j \alpha_j) x_i + \sum \beta_j$$

or, more compactly,

2.1.8
$$w = \alpha x_1 + \beta$$

where α and β depend only on the particular line considered.

Considering Figure II.1.2 once again, we see that, if the value of the coordinate x_1 is greater at Q than at P, then it must increase steadily from P to Q. Now, consider the objective function, w, which, as we have seen, can be expressed *along the line PQ* by (2.1.8). If $\alpha = 0$, then w is constant, equal to β, along this line. If $\alpha > 0$ then w increases whenever x_1 increases, and so it either increases steadily or decreases steadily along PQ. If $\alpha < 0$, then w decreases whenever x_1 increases, and once again, it either decreases or increases steadily along PQ. The important thing is that, along any given line l, the function w can do any one of these three things: remain constant, increase steadily, or decrease steadily. It cannot do anything else: it cannot, say, increase from P to R and then decrease from R to Q, or be constant from P to R and then increase from R to Q.

The importance of these observations can best be seen from the following theorem. We already know, from Chapter I, that the constraint set is a polyhedral set; we have seen how its vertices, or extreme points, can be found.

II.1.2. Theorem. The maximum of the objective function (2.1.1) over the constraint set, if such a maximum exists, will be attained at one of the extreme points of the constraint set.

We will not give a strict mathematical proof of this theorem here. The general idea of the proof runs as follows: first, for any point, P, we: shall let $w(P)$ denote the value of the objective function, w, at the point P. Now, suppose Q is the point that maximizes the function, i.e., Q lies in the constraint set, and, for any P in the constraint set, $w(P) \leq w(Q)$. If Q is an extreme point, the theorem is proved. If not, then there will be an extreme point, P, of the constraint set, such that the line PQ has points, such as R, which lie "beyond" Q (i.e., Q lies between P and R) but still inside the constraint set (Figure II.1.3).

FIGURE II.1.3 TheoremII.1.2

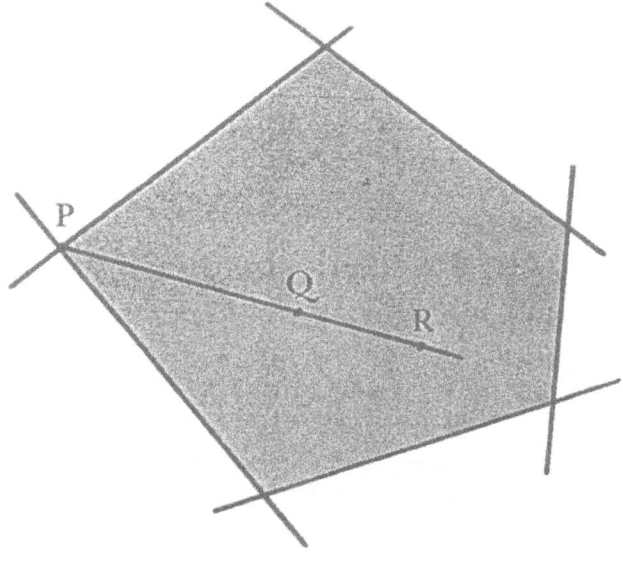

Suppose, now, that $w(P) < w(Q)$. By the linearity of w, we know that if w increases from P to Q, it continues to increase from Q to R, and so $w(Q) < w(R)$. But this contradicts the assumption that Q gave the maximum value of w. It follows that we cannot have $w(P) < w(Q)$; we must have $w(P) \geq w(Q)$. Thus, P also gives the maximum value for w. Since P is extreme, this proves the theorem.

In a sense, Theorem II.1.2 gives us a method for solving the linear program (2.1.1) to (2.1.3). Since the maximum is to be found at an extreme point, all that we have to do is consider all the extreme points, find the value of the objective function at each or these, and take the one that gives the best value. As there are only a finite number of extreme points, it is clear that this method will give us the solution in a finite number of steps

There are, unfortunately, two difficulties with this. The first is conceptual: we have, generally, no guarantee that the maximum exists, and, if it does not, Theorem II.1.2 will not tell us so. The second difficulty is practical: the number of extreme points, though finite, can be quite large, and finding them all might be beyond the scope even the largest computers in use today. Nevertheless we give examples of this method of solution of linear programs. We repeat that it is useful only in very small programs (i.e., those with only a few variables and constraints).

II.1.3 Example

Maximize
$w = 2x + y + 3z$

Subject to

$$x + 2y + z \geq 25 \quad (i_1)$$
$$3x + 2y + 2z \leq 30 \quad (i_2)$$
$$x \geq 0 \quad (i_3)$$
$$y \geq 0 \quad (i_4)$$
$$z \geq 0 \quad (i_5)$$

We shall, as in Chapter 1, look for the extreme points of the constraint set (It is for this reason that we have numbered the constraints.) As before, we take the constraints three at a time, and solve them as equations; taking constraints (i_1, i_2, i_3), we obtain the system

$$x + 2y + z = 25$$
$$3x + 2y + 2z = 30$$
$$x = 0$$

which has the solution (0,10,5). We check, then, that the remaining constraints are satisfied; since they are, we evaluate the objective function, which has the value of $2 \times 0 + 10 + 3 \times 5 = 25$ at this point.

Continuing in this way with each combination of three inequalities, we obtain Table II.1.2.

TABLE II.1.2

constraints	point	check	w
1,2,3	(0,10,5)	yes	25
1,2,4	(-20,0,45)	no	
1,2,5	(5/2,45/4,0)	yes	65/4
1,4,5	(25,0,0)	no	
2,3,4	(0,0,15)	yes	45
2,3,5	(0,15,0)	no	
1,3,4	(0,0,25)	no	
1,3,5	(0,25/2,0)	yes	25/2
2,4,5	(10,0,0)	yes	20
3,4,5	(0,0,0)	yes	0

We see from the table that, of the six extreme points of the constraint set, the point (0,0,15) gives the best value for the objective function, $w = 45$. We conclude that, if the program has a solution, it is this point.

The program does have a solution. It can be seen, multiplying the first constraint by 3, that $3x + 6y + 3z \leq 75$. Since the variables x, y, and z are non-negative, it follows that $w \leq 75$, i.e., in cannot become arbitrarily large. From this and the fact that the constraint set contains its boundary points, it can be shown that the function w will have a maximum. We have, thus, solved the program; the solution is

$$x = 0$$
$$y = 0$$
$$z = 15$$
$$w = 45$$

II.1.4. Example

Maximize
$$w = x + 2y$$
Subject to

$$-2x + y \leq 10 \quad (i_1)$$
$$x - 2y \leq 6 \quad (i_2)$$
$$x \geq 0 \quad (i_3)$$
$$y \geq 0 \quad (i_4)$$

Once again, we take the constraints two at a time and solve as equations. We obtain Table II.1.3.

TABLE II.1.3

constraints	point	check	w
1,2	(-8.7,-7.3)	no	
1,3	(0,10)	yes	20
1,4	(-5,0)	no	
2,3	(0,-3)	no	
2,4	(6,0)	yes	6
3,4	(0,0)	yes	0

We see from the table that, if the program has a solution, it must be the point (0, 10), which gives a value of 20 for w. Let us, however, look at Figure III.1.4, which shows the constraint set of the program. We see here that the constraint set is not bounded on the upper right-hand side; in fact, for any positive value of x, the point (x, x) lies in the constraint set. But this means that $w = x + 2y$ can be made arbitrarily large; the program fails to have a solution. In this case, our method of looking at the extreme points fails us.

PROBLEMS ON LINEAR PROGRAMMING

1.   Make a sketch of the constraint set in each of these problems

(a)   Maximize

$$3x + 5y$$

Subject to

$$x + 3y \leq 12$$

$$2x + y \leq 10$$
$$x + y \leq 4$$
$$x, y \geq 0$$

FIGURE II.1.4 Constraint set for Example II.1.4.

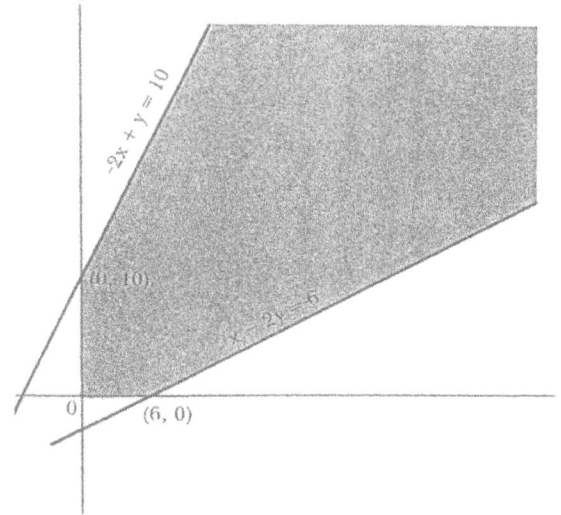

(b) Maximize

$$2x + y$$

Subject to

$$x + 3y \leq 12$$
$$2x + y \leq 10$$
$$x + y \leq 4$$

c) Minimize

$$3x + 2y$$

Subject to

$$x + 3y \geq 6$$
$$2x + y \geq 3$$
$$x \geq 0$$
$$y \geq 0$$

(d)   Maximize

$$3x - y$$

Subject to

$$x + 2y \geq 4$$
$$x + y \leq 12$$
$$2x + 4y \leq 30$$
$$x \geq 0$$
$$y \geq 0$$

2.   A dietitian must prepare a mixture of two foods A and B. Each unit of food A contains 10 gm. of protein and 20 gm. of carbohydrate, and costs 15¢. Each unit of food B contains 20 gm. of protein and 25 gm. of carbohydrate, and costs 20¢. What is the cheapest mixture that can be obtained, subject to the constraint that it must contain at least 500 gm. of protein and 800 gm. of carbohydrate?

3.   A nut company has 600 lb. of peanuts and 400 lb. of walnuts. It can sell the peanuts alone at 20¢ per lb.; it can also mix the peanuts and walnuts in a ratio of three parts peanuts and one part walnuts, or in a ratio of one part peanuts and two parts walnuts. The first mixture sells at 35¢ per lb. and the second at 50¢ per lb. How much of each mixture should it produce to maximize sales revenue?

4.   A food processing company has 1000 lb. of African coffee, 2000 lb. of Brazilian coffee, and 500 lb. of Colombian coffee. It produces two grades of coffee. Grade A is a mixture of equal parts of African and Brazilian coffees, and sells for 60¢ per lb. Grade B is a mixture of three parts Brazilian and one part Colombian coffee, and sells for 85¢ per lb. How much of each should it produce to maximize sales revenue?

2.   THE SIMPLEX ALGORITHM: SLACK VARIABLES

As we have seen from the examples in Section II.1, as a method of solving a linear program, the process of enumerating vertices is both impractical and uncertain. It is uncertain because we generally have no guarantee that the program does; indeed, have a solution; it may just possibly be unbounded, and the method of enumeration will not tell us when this happens. It is impractical because, for a large program, the process of enumerating the extreme points can be extremely long.

To take an example, a program with 6 variables and 10 constraints (including the non-negativity constraints) would require us to solve a total of 120 systems of 6 equations in 6 unknowns. Even if we could be certain of having the answer when we finished, the amount of work required would necessitate using a medium-sized computer to obtain the solution with reasonable speed. The trouble is that enumeration of the combinations of constraints, while it may be systematic is not to the point. A need arises, therefore, for a method of enumeration that will be both systematic *and* to the point (i.e., that will discard as many vertices as possible without more than passing glance, if that much). The *simplex algorithm* is such a method.

To understand more fully the point of the simplex algorithm, let us return to the geometric study of the program. As we have pointed out several times, the constraint set for a linear program will be a convex polyhedral set. We shall say that two vertices (extreme points) of this set are *adjacent extreme points* if they are joined by one of the edges of the polyhedron. In Figure II.2.1, the vertices A and B are adjacent, as are A and C, or C and D. The vertices B and C, however, are not adjacent; neither are D and E, nor C and F.

FIGURE II.2.1 Adjacent and non-adjacent vertices in a polyhedron.

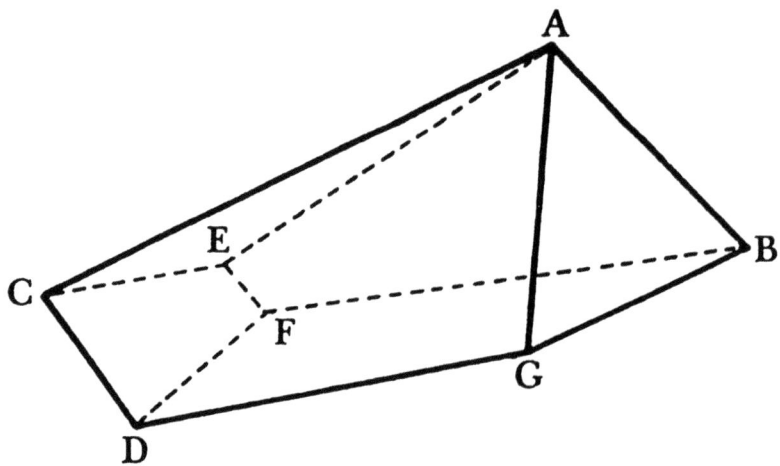

The simplex algorithm, as we shall see, is a procedure that allows us to go from one extreme point of the constraint set to an *adjacent* extreme point, in such a way that the value of the objective function is improved at each step, until we find the solution of the program (or until we see that it has no solution). As such, it is a very effective method of implicit enumeration of the extreme points - it automatically - discards a great many points before they are even considered.

The mathematical justification for the simplex algorithm is given by the following theorem:

II.2.1. Theorem. Let P be an extreme point of the constraint set of the program (2.1.1) through (2.1.3). Then, if P does not maximize the objective function w, there is an edge of the constraint set at P along which the objective function increases.

Again, we will not give a strict proof. The general idea of the proof, for a two-dimensional set, is as follows. Since P does not maximize w, there is a point in the constraint set, say Q, such that $w(Q) > w(P)$ (see Figure II.2.2). By the convexity of the constraint set and the linearity of the objective function we know that if R is on the line-segment PQ, then R is in the constraint set and $w(R) > w(P)$. Through R, a line may be drawn cutting the two edges that begin at P, at the points S and S'.

FIGURE II.2.2 Theorem II.2.1.

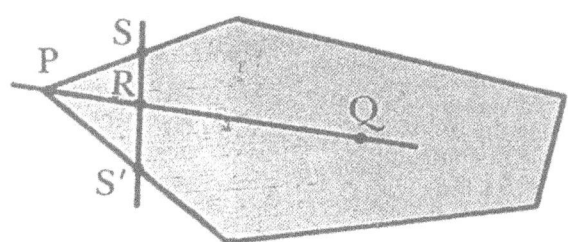

Now, if $w(S) \geq w(R)$, we have $w(S) > w(P)$, and so w increases along the edge PS. If, on the other hand, $w(S) < w(R)$, we must (by linearity) have $w(S') > w(R)$, and so $w(S') > w(P)$, so that w increases along the edge PS'. Thus w will increase along one, at least, of the two edges that begin at P.

For higher dimensions, the proof is similar, though somewhat greater care must be observed in the arguments. The theorem is true in any case.

Theorem II.2.1 tells us that the procedure just outlined, considering only extreme points adjacent to those that have already been obtained, will always bring us to the maximum if it exists. At the same time, if the objective function is not bounded, we will always find an edge along which the function increases without bound.

One of the difficulties in dealing with linear programs is that constraints of the type (2.1.2) are generally difficult to handle. On the other hand, equations are comparatively easy to handle, as are the non-negativity constraints (2.1.3). It follows that the program will become that much easier to solve if we can replace the constraints (2.1.2) by equations or byconstraints of the type (2.1.3). This can be effected by so-called *slack variables*.

Let us consider one of the constraints (2.1.2);

2.2.1 $$a_{i1}x_1 + a_{i2}x_2 + \ldots\ldots + a_{in}x_n \leq b_i$$

This can be restated in the form

$$a_{i1}x_1 + a_{i2}x_2 + \ldots\ldots + a_{in}x_n - b_i \leq 0$$

Since the left-hand side of this inequality is negative or zero, we can write

2.2.2 $$a_{i1}x_1 + a_{i2}x_2 + \ldots\ldots + a_{in}x_n - b_i = -u_i$$

where

2.2.3 $$u_i \geq 0$$

Thus we have replaced the single constraint (2.2.1) by the equation (2.2.2) and the non-negativity constraint (2.2.3). This is what we desired to do. Proceeding in this manner with each of the constraints (2.1.2), we find that the program (2.1.1) through (2.1.3) can be rewritten in the form

2.2.4 Maximize
$$w = c_1x_1 + c_2x_2 \ldots\ldots + c_nx_n$$

Subject to

2.2.5
$$a_{11}x_1 + a_{12}x_2 + \ldots\ldots + a_{1n}x_n - b_1 = -u_1$$
$$a_{21}x_1 + a_{22}x_2 + \ldots\ldots + a_{2n}x_n - b_2 = -u_2$$
$$\ldots\ldots\ldots\ldots\ldots\ldots\ldots\ldots\ldots\ldots\ldots\ldots\ldots$$
$$a_{m1}x_1 + a_{m2}x_2 + \ldots\ldots + a_{mn}x_n - b_m = -u_m$$

2.2.6
$$x_1, x_2, \ldots x_n \geq 0$$
$$u_1, u_2, \ldots u_m \geq 0$$

or, in matrix notation,

2.2.7 Maximize
$$w = c^t x$$

Subject to

2.2.8 $\qquad Ax - b = -u$
2.2.9 $\qquad x, u \geq 0$

3. THE SIMPLEX TABLEAU

We introduce next, a system of notation that simplifies the computational procedure for the simplex algorithm to a considerable extent. This is the *simplex tableau* or *schema*. In a sense, it is only a shorthand notation for a system of linear equations. The schema

2.3.1

x_1	x_2	$\ldots$	x_n	1	
a_{11}	a_{12}	$\ldots$	a_{1n}	$-b_1$	$= -u_1$
a_{21}	a_{22}	$\ldots$	a_{2n}	$-b_2$	$= -u_2$
$\ldots$	$\ldots$	$\ldots$	$\ldots$	$\ldots$	$\ldots$
$\ldots$	$\ldots$	$\ldots$	$\ldots$	$\ldots$	$\ldots$
$\ldots$	$\ldots$	$\ldots$	$\ldots$	$\ldots$	$\ldots$
a_{m1}	a_{m2}	$\ldots$	a_{mn}	$-b_m$	$= -u_m$
c_1	c_2	$\ldots$	c_n	0	$= w$

represents the equations (2.2.4) and (2.2.5). Generally speaking, each of the rows inside the box represents a linear equation, an equation obtained by multiplying each entry in the row by the corresponding entry in the top row, and then adding and setting this sum equal to the entry in the right-hand column (outside the box). Thus the first row represents the equation

$$a_{11}x_1 + a_{12}x_2 + \ldots + a_{1n}x_n - b_1 = -u_1$$

which is precisely the first of the equations (2.2.5), and so on for the other rows in the tableau. The bottom row in the tableau is nothing other than the definition of the objective function, w, according to (2.2.4).

Let us, for the present, ignore the non-negativity constraints (2.2.6). The equation constraints (2.2.5) are, of course, a system of m equations in the $m + n$ unknowns x_j, u_i. The system is "solved" for the u_i in terms of the x_j, but, as we know from elementary algebra, it is normally possible to solve such a system for any m variables in terms of the remaining n (though in practice there may be certain combinations of m variables for which this is not possible). What we are looking for now is a method of solving successively for different combinations of the variables.

To see how this is done, let us suppose that we have, at some time, solved for the variables $r_1, ..., r_m$ in terms of the variables $s_1, s_2, ..., s_n$. We have a tableau very similar to (2.3.1):

2.3.2

	s_1	s_2	....	s_n	1	
	a_{11}	a_{12}	....	a_{1n}	$-b_1$	$= -r_1$
	a_{21}	a_{22}	....	a_{2n}	$-b_2$	$= -r_2$
	....	....	....	....	....	....
	....	....	....	....	....	....
	....	....	....	....	....	....
	a_{m1}	a_{m2}	....	a_{mn}	$-b_m$	$= -r_m$
	c_1	c_2	....	c_n	0	$= w$

(where, of course, the entries are different from those in (2.3.1).) For such a tableau, the m variables $r_1, ... r_m$, for which we have solved, are called *basic variables*. The variables $s_1, ..., s_n$ in terms of which we have solved, are called *non-basic variables*. Let us suppose, next, that we want to interchange the roles of the basic variable r_i and the non-basic variable s_j; i.e., we wish to solve for m-1 of the r's, and s_j, in terms of n-1 of the s's, and r_i. Consider, then, the ith row of the tableau (2.3.2):

$$a_{j1}s_1 + a_{j2}s_2 + ... + a_{ij}s_j + ... + a_{in}s_n - b_i = -r_i$$

We can solve this for s_j, if $a_{ij} \neq 0$, obtaining

2.3.3 $\quad a_{j1}s_1/a_{ij} + a_{j2}s_2/a_{ij} + ... + r_i/a_{ij} + ... + a_{in}s_n/a_{ij} - b_i+/a_{ij} = -s_j$

We must now substitute this value of s_j in the remaining equations that (3.3.2) represents. Let us take any other equation, say the kth (in which $k \neq i$):

$$a_{k1}s_1 + a_{k2}s_2 + \ldots + a_{kj}s_j + \ldots + a_{kn}s_n - b_k = -r_k$$

Substituting (2.3.3), we obtain

2.3.4
$$(a_{k1} - a_{kj}a_{j1}/a_{ij}) s_1 + \ldots - (a_{kj}/a_{ij}) r_i + \ldots +$$
$$+ (a_{kn} - a_{kj}a_{jn}/a_{ij}) s_n - (b_k - b_j a_{kj}/a_{ij}) = -r_k$$

Equations (2.3.3) and (2.3.4) tell us, now, how the simplex tableau (2.3.2) is to be changed. We can summarize them as follows:

2.3.5. The variables r_j and s_i are interchanged.

2.3.6. The entry a_{ij}, called the *pivot* of the transformation, is replaced by its reciprocal, $1/a_{ij}$. 3;

2.3.7 The other entries a_{ih} or $-b_i$ in the pivot row, are replaced by a_{ih}/a_{ij} or $-b_i/a_{ij}$ respectively.

2.3.8 The other entries a_{kj}, or c_j in the pivot column, are replaced by $-a_{kj}/a_{ij}$ or $-c_j/a_{ij}$ respectively.

2.3.9 The other entries in the tableau, a_{kh}, $-b_k$, c_h **or** d, are replaced by $a_{kh} - a_{kj}a_{ih}/a_{ij}$, $-b_k + a_{kj}b_i/a_{ij}$, $c_h - a_{ih}c_j/a_{ij}$, and $d + b_i c_j/a_{ij}$ respectively.

A transformation such as that outlined in (2.3.5) is called a pivot step, or pivot transformation. We give the rules in schematic form

2.3.10

p	q
r	s

→

1/p	q/p
-r/p	s - qr/p

The schema (2.3.10) is to be interpreted as follows; p is the pivot, q is any other entry in the pivot row, r is any other entry in the pivot column, and s is the entry in the row of r and column of q. The arrow shows how the pivot step affects these entries. The pivot is replaced by its reciprocal; other entries in the pivot row are divided by the pivot; other entries in the pivot column are divided by the negative of the pivot. The entry s is replaced by $s-qr/p$ where p is the pivot and q and r are the two entries that form a rectangle with s and p (i.e., q is in the row of p and column of s, while r is in the column of p and row of s.

An example of such a pivot step follows.

II.3.1. Example. Solve the system
$$r = x + 2y - 3$$
$$s = 2x + 5y + 1$$
for x and y in terms of r and s.

There are actually two ways of doing this: one is by means of row operations on the matrix of coefficients; the other, by pivot steps. We shall use the latter method.

We form the tableau

x	y	1	
-1*	-2	3	= -r
-2	-5	-1	= -s

We wish to interchange the roles of x and y with those of r and s. This means we must carry out two pivot steps. In the first, we pivot on the starred entry to interchange the variables x and r, thus obtaining

r	y	1	
1/-1	-2/-1	3/-1	= -x
-(-2/-1)	-5 − (-2)(-2)/-1	-7	= -s

or, more concisely,

r	y	1	
-1	2	-3	= -x
-2	-1*	-7	= -s

This gives us x and s in terms of r and y. To interchange s and y, we pivot on the starred entry to obtain

	r	s	1	
	-5	2	-17	= -x
	2	-1	7	= -s

or, in equation form,

$$x = 5r - 2s + 17$$
$$y = 2r - s - 7$$

This could also have been done by means of row operations, as in Chapter II: we have the system

$$-x - 2y + r = -3$$
$$-2x - 5y + s = 1$$

which gives us the augmented matrix

-1	-2	1	0	-3
-2	-5	0	1	1

We multiply the first row by -1, and clear the first column:

1	2	-1	0	3
0	-1	-2	1	7

Now we multiply the second row by -1, and clear the second column, obtaining

1	0	-5	2	17
0	1	2	-1	-7

which gives us the desired solution. It may be checked that, as the row operations are carried out, the matrices obtained contain an identity matrix formed by two columns. The remaining columns are the same, except for a possible interchange of columns, as in the simplex tableaux obtained previously (the only difference being in the signs of the last column). In fact, the pivot steps are nothing other than a condensed format for the row operations.

II.3.2. Example. Interchange the roles of s and y in the tableau

x	y	1	
1	4	-1	= -r
5	-9*	2	= -s
6	-5	-3	= -t

In this case, the pivot is the entry -9, in the y-column and s-row. The new tableau will be

x	s	1	
29/9	4/9	-1/9	= -r
-5/9	-1/9	-2/9	= -y
29/9	-5/9	-37/9	= -t

PROBLEMS ON PIVOT TRANSFORMATIONS

1. In each of the following problems, solve the given system of equations for x, y, and z in terms of the other variables

(a)
$$3s + 2y - z + 4 = -x$$
$$s - 2y - z + 4 = -t$$
$$4s + 3y - 2z + 1 = -u$$

(b)
$$2x + 3y - 5 = -z$$
$$x + 2y - 3 = -s$$
$$3x + 7y + 1 = -t$$

(c)
$$x + 2s + 5z - 1 = -y$$
$$3x + s - 5z + 3 = -t$$
$$-x + 4s + 2z + 1 = -r$$

(d)
$$4x + 3y - 2z - 5 = -r$$
$$x + 2y - 3z + 3 = -s$$
$$2x + y + 2z + 6 = -t$$

$$3x + 2y + 6s + 1 = -r$$
$$x + y + 2s + 4 = -z$$
$$-2x - y + 3s - 5 = -t$$

(f)
$$5r + x - 2y + 4 = -z$$
$$3r + x + y - 6 = -s$$
$$r + 5x + 2y + 7 = -t$$

(g)
$$x + y + 3z - 2 = -s$$
$$2x + y - 2z + 4 = -t$$
$$5x + 2y - z + 6 = -u$$

4. THE SIMPLEX ALGORITHM: OBJECTIVES

Now that we have seen what a pivot step is, we have to decide what to do with it. We must remember that, essentially, it is nothing other than a method of rewriting the system of equations (2.2.5) to obtain a different but equivalent system. Let us suppose that, after one or several pivot steps, a tableau such as (2.3.2) is reached, which has the property that all the entries in the right-hand column, except possibly the bottom entry, are non-positive; that is, for each value of i, $b_i \geq 0$. Since (2.3.2) is just a "solution" of the equations (2.2.5) for the basic variables $r_1, \ldots, r_m$ in terms of the non-basic variables $s_1, \ldots, s_n$, we may assign values arbitrarily to the non-basic variables. We might let each of the non-basic variables equal zero. This gives us the values

2.4.1 $\qquad s_j = 0 \quad \text{for } j = 1, \ldots, n$

2.4.2 $\qquad r_i = b_i \quad \text{for } i = 1, \ldots, m$

as a particular solution to system (2.2.5). But we are assuming that $b_i \geq 0$. It follows that these values of the variables satisfy the non-negativity constraints (2.2.6) as well as the equations (2.2.5) and thus give us a point in the constraint set of the program (2.2.4) through (2.2.6).

Let us suppose, further, that each entry in the bottom row, except possibly the right-hand entry, is non-positive, i.e. for each j, $c_j \leq 0$. The bottom row of (2.3.2) says simply that

2.4.3 $\qquad w = c_1 s_1 + c_2 s_2 + \ldots\ldots\ldots + c_n s_n + \delta$

We know that a feasible point is obtained by letting each s_j equal zero. If we do so, we find that $w = \delta$, i.e., this feasible point gives a value of δ for the objective function. On the other hand, each term on the right-hand side of (2.4.3) except the last is the product of a non-positive constant c_j and a variable s_j that is non-negative at each feasible point. Thus, for each feasible point, the right side of (2.4.3) is the sum of n non-positive terms $c_j s_j$ and δ. It follows that we must have $w \leq \delta$ at all feasible points. But the point (2.3.1)-(2.3.2) gives a value of δ, and is thus the *solution* of the program: it gives the maximum of the objective function.

We know now what we want to accomplish by means of the pivot steps: we wish to make every entry in the right-hand column (except possibly the bottom entry) non-positive; we wish to make every entry in the bottom row (except possibly the right-hand entry) non-positive. Of course, if we wish to minimize the objective function, we will try to make entries in the bottom row (except the right-most one) non-negative. Except for this, there is no great difference between maximization and minimization problems. To see how this can best be done, we shall look at the geometric situation. First, however, we give some definitions.

II.4.1. Definition. A simplex tableau is associated with a point in $(m+n)$-dimensional space, namely the point obtained by setting the n non-basic variables equal to zero. Such a point is called a *basic point* of the program. If all the entries in the right-hand column (except possibly the bottom one) are non-positive, then the point is a *basic feasible point* (b.f.p.). Two basic points are said to be *adjacent* if the corresponding simplex tableaux can be obtained from each other by a single pivot step.

Geometrically, the situation is as follows: each of the planes $x_j = 0$ or $u_i = 0$ is one of the bounding hyperplanes of the constraint set. The intersection of n of these determines a basic point of the program. If the remaining constraints are all satisfied, we have a vertex of the constraint set, so that a b.f.p. is simply one of the vertices of the constraint set. An edge of this set is determined by setting $n-1$ of the variables equal to zero, so that two vertices will lie on a common edge if they have $n-1$ of their non-basic variables in common, i.e., if their tableaux can be obtained from each other by a single pivot step. The idea of adjacent b.f.p.'s corresponds to the geometric idea of adjacent vertices.

5. THE SIMPLEX ALGORITHM: CHOICE OF PIVOTS

The process of solving a linear program through the simplex method generally can be thought of as consisting of two stages. In Stage I, the entries in the right-hand column are made non-negative to obtain a basic feasible point. In Stage II, while care is exercised to maintain a b.f.p. at all times, the entries in the bottom row are made non-positive (or non-negative in case of a minimization problem) to obtain the solution. The two stages have to be considered separately since the objectives in the two cases are distinct. Let us consider Stage II first; the rules for dealing with Stage I can be adapted from those for Stage II.

For Stage II, Figure II.5.1 gives a two-dimensional illustration of what we are trying to do. The vertex A lies at the intersection of $x = 0$ and $r = 0$; i.e., its non-basic variables are r and x. From this point, suppose we see that the objective function increases along the edge AB, i.e., by letting x increase while r remains at zero. We improve the value of the function, then, if we interchange x with one of the basic variables y, s, t. But which one? Clearly not y, since this would cause x to decrease, giving the point D. Not t either, since this would cause x to increase too much; it would give us the point C, at which s is negative. We conclude that we should interchange x and s since, of those interchanges that cause x to increase, this is the one that causes the smallest increase. We thus obtain the adjacent vertex, B, and continue from B in the same manner.

FIGURE II.5.1 Movement from A to B

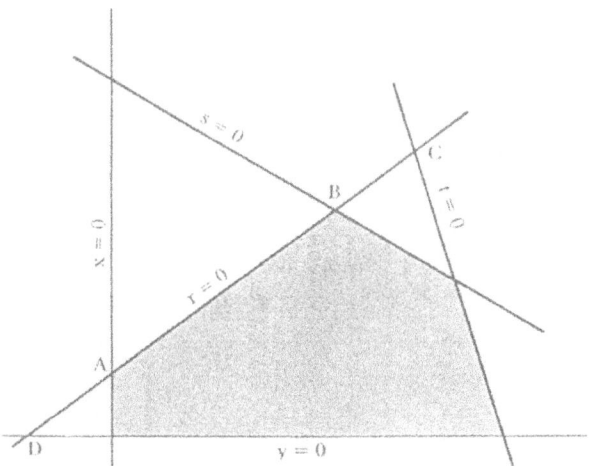

Accordingly, we obtain the following rules:

II.5.1. Rules for Choosing a Pivot (Stage II). Let c_j be any positive entry (other than the right-hand entry) in the bottom row of the tableau. This gives us the pivot column. For each positive entry a_{ij} in this column, form the quotient b_i/a_{ij} (obtained by dividing a_{ij} into the negative of the corresponding entry in the right-hand column). Let b_k/a_{kj} be the smallest of these quotients. Then a_{kj} will be the pivot. (If there is a tie for the smallest quotient, the pivot may be chosen arbitrarily from among those that tie.)

The rules II.5.1 are for a maximization problem. If we are required to minimize the objective function, the procedure is changed only in that we take c_j to be a negative entry; otherwise, the choice of pivot is the same.

We shall not, at this moment, prove that the rules II.5.1 work in the sense that they will eventually give us the solution of the program if such a solution exists. We give, instead, a numerical example.

II.5.2. Example. A company produces three types of toys, A, B, and C, by using three machines called a forge, a press, and a painter. Each toy of type A must spend 4 hr. in the forge, 2 hr. in the press, and 1 hr. in the painter, and yields a profit of $5. Each toy of type B must spend 2 hr. in the forge, 2 hr. in the press, and 3 hr. in the painter and yields a profit of $3. Each toy of type C spends 4 hr. in the forge, 3 hr. in the press, and 2 hr. in the painter and yields a profit of $4. In turn, the forge may be used up to 80 hr. per week; the press up to 50 hr., and the painter up to 40 hr. How many of each type of toy should be made in order to maximize the Company's profits?

FIGURE II.5.2 Constraint set for Example II.5.2. Note that each bounding plane corresponds to one of the variables

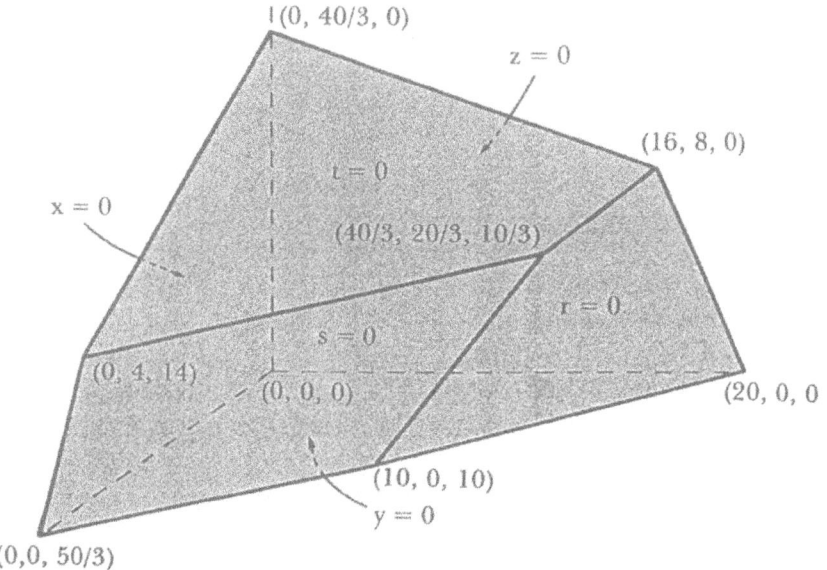

Letting x, y, and z be the amounts produced of toys A, B, and C respectively, we obtain the linear program:

Maximize
$$5x + 3y + 4z = w$$
Subject to
$$4x + 2y + 4z \leq 80$$
$$2x + 2y + 3z \leq 50$$
$$x + 3y + 2z \leq 40$$
$$x, y, z \geq 0$$

Figure II.5.2 shows the constraint set for this problem. Introducing the slack variables r, s, and t, we obtain the tableau

x	y	z	1	
4*	2	4	-80	= -r
2	2	3	-50	= -s
1	3	2	-40	= -t
5	3	4	0	= w

FIGURE II.5.3 At the vertex (0,0,0), the function increases along each edge.

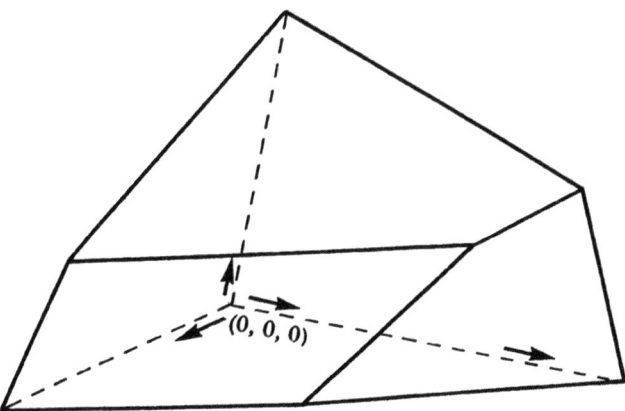

We see that we have a basic feasible point, since the entries in the right-hand column are negative. On the other hand, there are three positive entries in the bottom row (see also Figure II.5.3). Let us choose the first column, which has one

of these positive entries, as the pivot column. According to the rules II.5.1, we divide each of the positive entries in this column into the corresponding b_i.

The corresponding quotients b_i/a_{ij} are then

$$b_1/a_{11} = 80/4 = 20$$

$$b_2/a_{21} = 50/2 = 25$$

$$b_3/a_{31} = 40/1 = 40$$

The first row gives the smallest quotient; the entry 4 (starred) will be the pivot. This gives us the new tableau

r	y	z	1	
1/4	1/2	1	-20	= -x
-1/2	1	1	-10	= -s
-1/4	5/2*	1	-20	= -t
-5/4	1/2	-1	100	= w

FIGURE II.5.4 At (20,0,0), the function increases along one of the three edges.

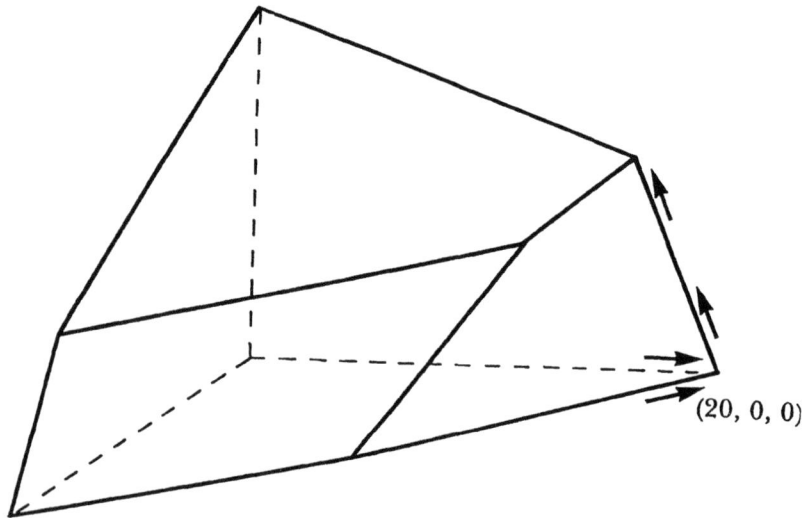

This tableau gives an improved b.f.p.: the profit here is $100. It is still not the maximum, since there is a positive entry in the bottom row, second column (see also Figure II.5.4). Once again, we consider the quotients b_i/a_{ij} for each positive a_{ij} in the second column. These are, respectively, $20/(1/2) = 40$, $10/1 = 10$, and $20/(5/2) = 8$. Since the starred entry, 5/2, gives us the smallest quotient, we use it as pivot. We then obtain the tableau

r	t	z	1	
3/10	-1/5	4/5	-16	= -x
-2/5	-2/5	3/5	-2	= -s
-1/10	2/5	2/5	-8	= -y
-6/5	-1/5	-6/5	104	= w

(see Figure II.5.5). This gives us the solution: it is obtained by setting the non-basic variables equal to zero, and so

$$x = 16$$
$$y = 8$$
$$z = 0$$
$$w = 104$$

FIGURE II.5.5 At (16,8,0), the function decreases along each edge.

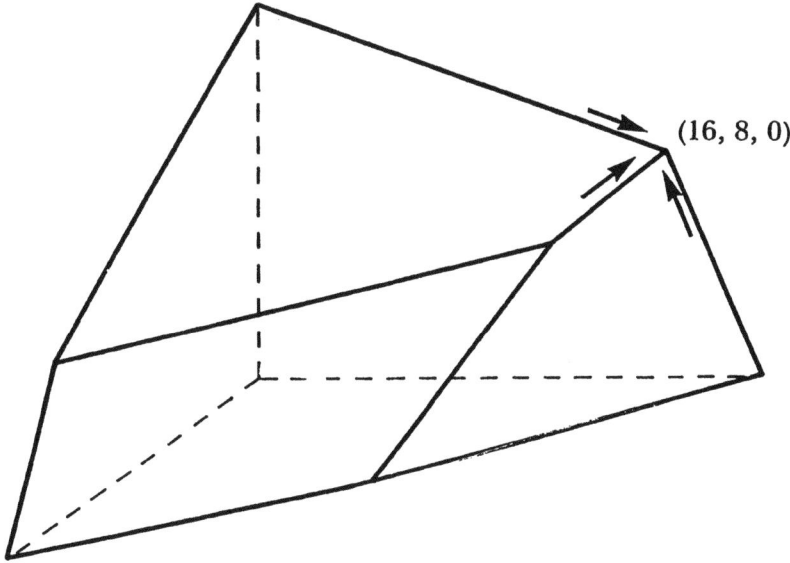

The company should produce 16 type A toys and 8 type B toys every week. Type C toys will not be produced. It may be checked directly that the constraints are satisfied; it is not possible, at least yet, to check that this is indeed the solution. Note that the press is idle 2 hr a week; this corresponds to the value $s = 2$ for the related slack variable.

II.5.3. Example
Maximize
$$w = 2x + y + 3z$$

Subject to
$$x + 2y + z \leq 25$$
$$3x + 2y + 2z \leq 30$$
$$x, y, z \geq 0$$

This is, of course, the same as Example II.1.3. We proceed to solve it, now, by the simplex algorithm. Introducing slack variables, we have the b.f.p.

x	y	z	1	
1	2	1	-25	$= -r$
3	2	2*	-30	$= -s$
2	1	3	0	$= w$

We may choose any one of the columns as the pivot column. Say we choose the third column. The corresponding quotients are $25/1 = 25$ and $30/2=15$. Since 15 is the smaller quotient, we choose the starred entry as pivot. We obtain

x	y	s	1	
-1/2	1	-1/2	-10	$= -r$
3/2	1	1/2	-15	$= -z$
-5/2	-2	-3/2	45	$= w$

which gives the solution $x = 0, y = 0, z = 15, w = 45$. This is, of course, the same as was obtained by other means earlier in the chapter.

6. THE SIMPLEX ALGORITHM: STAGE I

In Stage I, the situation is, as we mentioned, somewhat different. The general rule is to take the several entries $-b_i$ in the right-hand column and work on them, one at a time, until all have become non-positive. Some care must be observed, of course, that while one $-b_i$ becomes non-positive, others do not simultaneously become positive. The method for working on a positive entry $-b_i$, then, is as follows: if it can be made non-positive immediately without disturbing non-positivity of the other entries in the column, we do so. If not, we treat the $-b_i$ as a "secondary objective≅: we try to decrease it as much as possible while keeping the other entries in the column non-positive. We have the following rules.

II.6.1. Rules for Choosing a Pivot (Stage I). Let $-b_l$ be the lowest positive entry (other than the bottom entry) in the right-hand column. Let a_{lj} be any negative entry in this row; this gives us the pivot column. Now, for a_{lj}, and for each positive entry a_{ij} below a_{lj} in this column, form the quotient b_i/a_{ij}. Let b_k/a_{kj} be the smallest of these quotients. Then a_{kl} is the pivot. (Once again, in case of ties, any one of those entries that tie to give the smallest quotient may be chosen as pivot.)

We shall leave the proof that these rules work, i.e., that they eventually give us a b.f.p., until later. We are now in a position to solve Example III.1.1, which follows.

II.6.2. Example
Minimize
$$w = 0.60x + 1.25y + 0.50z$$
Subject to
$$3x + 5y + 3z \geq 20$$
$$2x + 6y + 4z \geq 18$$
$$6x + 2y + 5z \geq 30$$
$$x, y, z \geq 0$$

Introducing slack variables, we have the tableau

x	y	z	1	
-3	-5	-3	20	= -r
-2	-6	-4	18	= -s
-6	-2	-5*	30	= -t
3/5	5/4	½	0	= w

which does not represent a b.f.p., since there are several positive entries in the right-hand column. The lowest such entry is in the third row. We take a negative entry in this row, say the -5 in the third column. According to the rules II.3.5, we must consider this entry and any positive entry below it except that in the bottom row. This reduces the choice to no choice at all, of course; and the -5 will be the pivot. We then obtain

x	y	t	1	
3/5	-19/5	-3/5*	2	= -r
14/5	-22/5	-4/5	-6	= -s
6/5	2/5	-1/5	-6	= -z
0	21/20	1/10	3	= w

which is still not a b.f.p. We now work on the top row: there are two negative entries here, and either one can be used to obtain the pivot column. Let us take the third column. There are no positive entries in this column, other than the bottom entry, so that the starred entry, -3/5, will be the pivot. We then obtain

x	y	r	1	
-1	19/3	-5/3	-10/3	= -t
2	2/3	-4/3	-26/3	= -s
1	5/3	-1/3	-20/3	= -z
1/10	5/12	1/6	10/3	= w

which is not only a b.f.p., but actually the solution since all the entries in the bottom row are positive (and we want to minimize the function) We have then

$$x = 0$$
$$y = 0$$
$$z = 20/3$$
$$w = 10/3$$

and the mixture desired in Example II.1.1 consists of 20/3 units of food of type C, and nothing else, for a total cost of $3.33.

II.6.3. Example

Maximize
$$w = 2x + y + z$$
Subject to
$$x - 10y - 4z \leq -20$$
$$3x + y + z \leq 3$$
$$x, y, z \geq 0$$

We obtain, in this case, the tableau

x	y	z	1	
1	-10	-4	20	= -r
3	1	1*	-3	= -s
1	2	1	0	= w

There is a positive entry at the top of the right-hand column. We can choose the pivot column to be either the second or third column, since these are the negative entries in the top row. Let us choose the third. The pivot is then chosen from between the negative entry -4 and the positive entry below it (the bottom row does not count). The corresponding quotients are -20/-4 = 5 and 3/1 = 3. Since 3 is smaller, we choose the starred entry as pivot. This gives us

x	y	s	1	
13	-6*	4	8	= -r
3	1	1	-3	= -z
-2	1	-1	3	= w

which is still not a b.f.p. Note how the right-hand entry in the top row has been decreased, while the right-hand entry in the second row remains non-positive.

The only negative entry in the first row is in the second column, which is, therefore, the pivot column. The choice is between the negative entry in the top row and the positive entry below it; the corresponding quotients are -8/-6 =4/3 and 3/1 = 3, and the pivot is the starred entry. We then get the tableau

x	r	s	1	
-13/6	-1/6	-2/3	-4/3	= -y
31/6	1/6*	5/3	-5/3	= -z
1/6	1/6	-1/3	13/3	= w

which is a b.f.p., but not the solution, since there are two positive entries in the bottom row. Let us choose the second column as the pivot column; the pivot will be the only positive entry in this column (other than the bottom entry) which is starred. We then get the tableau

x	z	s	1	
3	1	1	-3	= -y
31	6	10	-10	= -r
-5	-1	-2	6	= w

which gives us the solution: it is

$$x = 0$$
$$y = 3$$
$$z = 0$$
$$w = 6$$

7. THE SIMPLEX ALGORITHM: PROOF OF CONVERGENCE

Now that we have illustrated the two sets of rules, II.5.3 and II.6.1, for choosing pivots, we shall show that they actually work in the sense of giving us the solution if this exists and telling us that there is no solution if such is the case. We shall make, at first, an assumption of *non-degeneracy:* no more than n of the bounding hyperplanes of the original program (2.1.1) to (2.1.3) pass through a single point. This is the general case and will "almost always" happen, unless the program is of a very special type. In practice this means that there will be no zeros in the right-hand column (except, possibly, in the bottom row).

II.7.1. Proof for Stage II.

In Stage II, the constant term in the pivot row, $-b_k$, is replaced by

2.7.1 $-b_k' = -b_k/a_{kj}$

By hypothesis, $-b_k \leq 0$, and $a_{kj} > 0$. Hence $-b_k' \leq 0$.

The other terms $-b_i$ in the right-band column are replaced by

2.7.2 $\qquad -b_i' = -b_i + b_k a_{ij}/a_{kj}$

We consider two possibilities: $a_{ij} \le 0$ and $a_{ij} > 0$. If $a_{ij} \le 0$ then, since $b_k \ge 0$ and $a_{kj} > 0$, it will follow that the second term on the right-hand side of (2.7.2) is non-positive. In this case, $-b_i' \le -b_i$. But, by hypothesis, $-b_i \le 0$. Hence $-b_i' \le 0$.

If $a_{ij} > 0$, then (2.7.2) can be rewritten

2.7.3 $\qquad -b_i' = a_{ij}(b_k/a_{kj} - b_i/a_{ij})$

Now, $a_{ij} > 0$, which means that it was among the terms considered as possible pivots. The entry a_{kj} was chosen as pivot, which means (by rules II.5.1) that

$$b_k/a_{kj} \le b_i/a_{ij},$$

But this means that the parenthesis on the right side of (2.7.3) is non-positive. Since $a_{ij} > 0$, it follows that $-b_i' \le 0$.

Finally, the term δ, in the lower right-hand corner of the tableau, is replaced by

2.7.4 $\qquad \delta' = \delta + b_i c_j / a_{ij}$

Now, according to the rules, c_j and a_{ij} are both positive. By hypothesis, $b_i \ge 0$. Under the non-degeneracy assumption, however, we cannot have $b_i = 0$ so that $b_i > 0$. This means that the second term on the right side of (2.7.4) is positive, and so $\delta' > \delta$.

We have seen, then, that following the rules II.5.1 will give us a new tableau that represents a b.f.p. (since all the $-b_i'$ are non-positive). Moreover, this new b.f.p. will (under the non-degeneracy assumption) give an improved value for the objective function. Thus, every time that we follow the rules we obtain a new vertex that is an improvement over the previous one. This means that we cannot obtain the same vertex twice. As there are only a finite number of vertices, it follows that we must eventually get a tableau at which the rules II.5.1 cannot be followed.

Suppose, then, that it is impossible to follow the rules. This may be for either of two reasons. One is that there are no positive entries (other than the right-hand entry) in the bottom row. But this means that we have the solution.

The other possibility is that there may be a column (other than the right-hand column) with a positive entry in the bottom row, and no other positive entries, i.e., for some j, $c_j > 0$, but $a_{ij} \leq 0$ for each i. If we let all the non-basic variables, except s_j, vanish, we have

$$r_i = b_i - a_{ij} s_j \quad \text{for each } i$$
$$w = \delta + c_j s_j$$

Now, we can see that, for any positive value of s_j, we will have $r_i \geq b_i$ (since $a_{ij} \leq 0$). But $b_i \geq 0$ (by hypothesis, we have a b.f.p.) so $r_i \geq 0$. On the other hand, since $c_j > 0$, it follows that we can make w as large as desired simply by making s_j large enough. But this means that the program has no solution: it is unbounded. We have seen that starting From any b.f.p., the rules II.5.1 will eventually *either* give us the solution to the program *or* tell us that the program is unbounded. An example of the latter possibility follows.

II.7.2. Example.

Maximize
$$w = 2x + y + z$$

Subject to
$$x - 2y - 4z \leq -20$$
$$3x - 5y + z \leq 3$$
$$x, y, z \geq 0$$

We obtain, in this case, the tableau

x	y	z	1	
1	-2	-4	20	= -r
3	-5	1*	-3	= -s
1	2	1	0	= w

We have a positive entry in the right-hand column. Taking the first column as our pivot column, the rules II.6.1 tell us to pivot on the starred entry, giving us the tableau

x	y	s	1	
13	-22*	4	8	= -r
3	-5	1	-3	= -z
-2	7	-1	3	= w

which is still not a b.f.p. We pivot on the starred entry (which is the only negative entry in the column) obtaining

x	r	s	1	
-13/22	-1/22	-2/11	-4/11	= -y
1/22*	-5/22	1/11	-53/11	= -z
47/22	7/22	3/11	61/11	= w

which is a b.f.p., but not the solution, as there are several positive entries in the bottom row. If we pivot on the starred entry, we obtain the tableau

z	r	s	1	
13	-3	1	-63	= -y
22	-5	2	-106	= -x
-47	11	-4	232	= w

This tableau is not the solution, as there is still a positive entry in the bottom row, second column. There are, however, no other positive entries in the second column, so the rules II.5.1 cannot be followed. We conclude that the program does not have a solution: it is unbounded. In fact, we may check that, for any positive value of r, the point

$$x = 106 + 5r$$
$$y = 63 + 3r$$
$$z = 0$$
$$w = 232 + 11r$$

satisfies the constraints. By making r large enough, w can be made as large as desired. If, for example, we want w to be greater than 1,000,000, we simply let $r = 100,000$; then

$$x = 500,106$$
$$y = 300,063$$
$$z = 0$$
$$w = 1,100,232$$

is a feasible point of the program, as may be checked directly.

We return to the rules II.6.1. Once again we make the nondegeneracy assumption: no zeros appear in the right-hand column (except possibly in the bottom row).

II.7.3. Proof for Stage I.

In Stage I, the constant term in the pivot row, $-b_k$, and the other terms in the right-hand column, $-b_i$, are replaced by $-b_k'$ and $-b_i'$, respectively, according to equations (2.7.1) and (2.7.2). It can be shown, exactly as in II.7.1, that, for $i > 1$, $-b_i \leq 0$ (i.e., the entries in the right-hand column below the ith row remain non-negative.)

As for the entry $-b_l$, it is replaced by $-b_l'$ which will be given by (2.7.1) or (2.7.2), depending on whether the pivot is in the lth row ($k=l$) or in another row ($k \neq l$).

In case the pivot is in the lth row we have

2.7.5 $-b_l' = -b_l / a_{il}$

now, by assumption, b_l and a_{il} are both negative. This means that $-b_l' < 0$.

If the pivot is in another row, we will have

2.7.6 $-b_l' = -b_l + b_k a_{il} / a_{ik}$

By the rules, $a_{il} < 0$ while $a_{ik} > 0$. By the non-degeneracy assumption, $b_k > 0$, since $k > l$. It follows that the second term on the right side of (2.7.6) is negative and so $-b_l' < -b_l$.

We see, thus, that, by following the rules II.6.1, we obtain a new tableau that preserves the non-positivity of the terms in the right-hand column below the lth row. Under the non-degeneracy assumption, the term $-b_l$ (the lowest positive entry in the right-hand column) will be decreased. Using once again the fact that any program can give only a finite number of distinct tableaux, we conclude that, so long as we follow the rules II.6.1, the term $-b_l$ will eventually become negative. We then work on the lowest remaining positive term in the right-hand column, and it follows that, eventually, *either* a b.f.p. will be reached *or* a tableau will appear at which it is impossible to follow rules II.6.1.

Suppose then, that we find it impossible to follow the rules II.6.1, although we do not yet have a b.f.p. This will happen if a row has a positive entry in the right-hand column, and only non-negative entries otherwise, i.e., if for some i, $-b_i > 0$, and $a_{ij} \geq 0$ for all j. The ith row then represents the equation

2.7.7
$$r_i = b_i - (a_{i1}s_1 + a_{i2}s_2 + \ldots + a_{in}s_n)$$

Now, each term inside the parenthesis in the right side of (2.7.7) is the product of a non-negative constant a_{ij} and a non-negative variable s_j. It follows that the expression inside the parenthesis is non-negative, and so $r_i \leq b_i$. But $b_i < 0$ and so $r_i < 0$. However, r_i must be non-negative. We have, thus, a contradiction. This means that the program is infeasible: the constraints cannot be satisfied.

We see, then, that the rules for Stage I will either lead to a b.f.p. or tell us that the program is infeasible. An example of this latter possibility is:

II.7.4. Example
Maximize
$$w = x - 4y$$
Subject to
$$x + 3y \geq 2$$
$$2x + 5y \leq 1$$
$$x + y \leq 4$$
$$x, y \geq 0$$

We have here the tableau

x	y	1	
-1	-3	2	= $-r$
2	5*	-1	= $-s$
1	1	-4	= $-t$
1	-4	0	= w

which has a positive entry in the right-hand column, top row. There are two negative entries in this row: let us choose the second column as the pivot column. The negative entry in the top row and both positive entries below it have to be considered in choosing a pivot. The corresponding quotients are 2/3, 1/5, and 4. The smallest is in the second row; we pivot, therefore, on the starred entry to obtain the tableau

x	s	1	
1/5	3/5	7/5	$= -r$
2/5	1/5	-1/5	$= -y$
3/5	-1/5	-19/5	$= -t$
13/5	4/5	-4/5	$= w$

This is not a b.f.p., since there is still a positive entry (albeit smaller) in the right-hand column. In fact, however, all the entries in the top row are positive. We conclude that the program is infeasible.

PROBLEMS ON TRANSPORTATION AND ASSIGNMENT

1. Solve the following transportation problems. In each case, the matrix $C = (c_{ij})$ is the cost matrix, while $\mathbf{a} = (a_1, \ldots, a_n)$ and $\mathbf{b} = (b_1, \ldots, b_n)$ are the availability and requirement vectors, respectively. (In cases, in which the total availability is greater than the total requirement, introduce a *fictitious destination* to receive the remaining goods, with 0 costs from each warehouse to this destination.)

(a)

$$C = \begin{array}{|c|c|c|} \hline 3 & 1 & 4 \\ \hline 1 & 5 & 9 \\ \hline 2 & 6 & 5 \\ \hline \end{array} \qquad \mathbf{a} = \begin{array}{|c|} \hline 100 \\ \hline 150 \\ \hline 120 \\ \hline \end{array}$$

$$\mathbf{b} = \begin{array}{|c|c|c|} \hline 140 & 90 & 140 \\ \hline \end{array}$$

(b)

$$C = \begin{array}{|c|c|c|c|} \hline 3 & 5 & 8 & 9 \\ \hline 7 & 9 & 3 & 2 \\ \hline 3 & 8 & 4 & 6 \\ \hline \end{array} \qquad \mathbf{a} = \begin{array}{|c|} \hline 50 \\ \hline 70 \\ \hline 40 \\ \hline \end{array}$$

$$\mathbf{b} = \begin{array}{|c|c|c|c|} \hline 30 & 45 & 30 & 55 \\ \hline \end{array}$$

(c)

$$C = \begin{bmatrix} 1 & 4 & 5 & 1 & 2 \\ 2 & 6 & 1 & 3 & 5 \\ 1 & 3 & 0 & 1 & 1 \end{bmatrix} \quad a = \begin{bmatrix} 80 \\ 80 \\ 40 \end{bmatrix}$$

$$b = \begin{bmatrix} 50 & 50 & 20 & 35 & 45 \end{bmatrix}$$

(d)

$$C = \begin{bmatrix} 1 & 4 & 1 & 5 & 2 \\ 2 & 6 & 3 & 4 & 1 \\ 5 & 8 & 5 & 6 & 4 \end{bmatrix} \quad a = \begin{bmatrix} 90 \\ 90 \\ 30 \end{bmatrix}$$

$$b = \begin{bmatrix} 40 & 60 & 25 & 35 & 40 \end{bmatrix}$$

(e)

$$C = \begin{bmatrix} 2 & 6 & 5 & 3 & 5 \\ 2 & 7 & 1 & 8 & 2 \\ 1 & 4 & 1 & 4 & 2 \\ 1 & 7 & 3 & 2 & 3 \end{bmatrix} \quad a = \begin{bmatrix} 50 \\ 50 \\ 50 \\ 50 \end{bmatrix}$$

$$b = \begin{bmatrix} 40 & 40 & 40 & 40 & 40 \end{bmatrix}$$

2. Formulate the duals of the foregoing problems (a) through (e). What heuristic interpretation can be given to them?

3. Solve the following assignment problems. In each problem, each row represents a candidate, and each column represents a position. (If there are more candidates than positions, introduce *fictitious positions,* representing no job, with all entries 0 in the corresponding columns.) What salaries should be paid to the employees in each case?

(a)

7	2	8	4	3
5	1	6	3	3
6	3	7	4	5
5	2	5	2	4
3	0	4	3	1

(b)

1	4	1	5	9	2	6
5	3	5	8	9	7	9
3	2	3	8	4	6	2
6	4	3	3	8	3	2
7	9	3	7	5	1	0
6	0	3	9	9	2	7
1	8	2	8	5	3	4

(c)

1	2	8	3	5
2	6	7	5	2
4	9	6	3	3
2	4	8	5	5
1	3	5	4	1
3	8	8	5	2

(d)

2	6	5	8	4	3
1	3	5	7	9	2
2	5	7	8	1	4
3	6	9	9	2	5
1	5	4	6	2	5
4	7	8	3	1	6
2	3	6	3	1	0

8. EQUATION CONSTRAINTS

In some cases, some of the constraints in a program may equations instead of inequalities. As we mentioned earlier, an equation may be replaced by two opposite inequalities, so that it is always possible to reduce the program to standard form. On the other hand, it should be clear that if we replace an equation by two inequalities, and then introduce two slack variables to make each of those inequalities into an equation, we are merely multiplying work for ourselves. It is better to leave the equations as they are: there is then, of course, no slack variable corresponding to the constraint. The technique is slightly different, though the same general rules apply. The following example shows this.

II.8.1. Example. A steel company has two warehouses, W_1 and W_2, and three distributors, D_1, D_2, and D_3 The two warehouses hold, respectively, 50,000 and 60,000 tons of steel. Of these, 30,000 tons must be shipped to D_1, 40,000 tons to D_2, and 40,000 tons to D_3. The transportation costs, in dollars per ton, between warehouses and distributors, are given in Table II.8.1.

TABLE II.8.1

From Warehouse	To Distributor		
	D_1	D_2	D_3
W_1	1	4	1
W_2	5	9	2

The required amounts must be transported in such a way as to minimize the total costs.

Letting x_{ij} represent the amount to be shipped from warehouse W_i to distributor D_j, we have the program

Minimize
$$w = x_{11} + 4x_{12} + x_{13} + 5x_{21} + 9x_{22} + 2x_{23}$$
Subject to
$$\begin{aligned}
x_{11} + x_{12} + x_{13} &= 50 \\
x_{21} + x_{22} + x_{23} &= 60 \\
x_{11} \quad\quad\quad\quad + x_{21} \quad\quad\quad\quad &= 30 \\
x_{12} \quad\quad\quad\quad + x_{22} \quad\quad &= 40 \\
x_{13} \quad\quad\quad\quad + x_{23} &= 40 \\
x_{ij} &\geq 0
\end{aligned}$$

where, for example, the first constraint means that the amounts shipped from W_1 to the three distributors must add up to 50, and so on for the other constraints (we are measuring in units of a thousand tons). Of the five equation constraints, one can be dispensed with: the fifth constraint is equal to the sum of the first two, minus the sum of the third and fourth equations. (This is, of course, because the amounts available at the warehouses are equal to the total requirements by the distributors. If D_1 and D_2 receive what they require, then the amount left for D_3 is exactly what is required there.) Discarding the fifth equation, we get the tableau

x_{11}	x_{12}	x_{13}	x_{21}	x_{22}	x_{23}	1	
1	1	1	0	0	0	-50	= 0
0	0	0	1	1	1*	-60	= 0
1	0	0	1	0	0	-30	= 0
0	1	0	0	1	0	-40	= 0
1	4	1	5	9	2	0	= w

It is clear that zeros have no business in the space reserved for basic variables. To remove them, we pivot in the usual manner, say on the starred entry. This will bring one of the zeros into the row of non-basic variables:

x_{11}	x_{12}	x_{13}	x_{21}	x_{22}	~~0~~	1	
1	1	1	0	0	~~0~~	-50	= 0
0	0	0	1	1	~~1~~	-60	= $-x_{23}$
1	0	0	1	0	~~0~~	-30	= 0
0	1	0	0	1*	~~0~~	-40	= 0
1	4	1	3	7	~~-2~~	120	= w

We cross out the column corresponding to the zero, since the entries in that column would merely be multiplied by zero. Pivoting on the next starred entry, we get

x_{11}	x_{12}	x_{13}	x_{21}	~~0~~	1	
1	1	1	0	~~0~~	-50	= 0
0	-1	0	1	~~1~~	-20	= $-x_{23}$
1	0	0	1*	~~0~~	-30	= 0
□0	1	0	0	~~1~~	-40	= $-x_{22}$
1	-3	1	3	~~-7~~	400	= w

Again, we cross out the column corresponding to zero, and pivot on the starred entry:

x_{11}	x_{12}	x_{13}	~~0~~	1	
1	1	1*	~~0~~	-50	= 0
-1	-1	0	~~1~~	10	= $-x_{23}$
1	0	0	~~1~~	-30	= $-x_{21}$
0	1	0	~~0~~	-40	= $-x_{22}$
-2	-3	1	~~3~~	490	= w

We repeat the procedure:

x_{11}	x_{12}	~~0~~	1	
1	1	~~1~~	-50	= $-x_{13}$
-1	-1*	~~0~~	10	= $-x_{23}$
1	0	~~0~~	-30	= $-x_{21}$
0	1	~~0~~	-40	= $-x_{22}$
-3	-4	~~1~~	540	= w

Having reduced the tableau to the desired form, we now proceed to solve the problem by applying the rules II.5.1 and II.6.1. We show the pivots, in each case, by asterisks:

x_{11}	x_{23}	1	
0	1	-40	= $-x_{13}$
1	-1	-10	= $-x_{12}$
1	0	-30	= $-x_{21}$
-1	1*	-30	= $-x_{22}$
1	-4	500	= w

x_{11}	x_{22}	1	
1*	-1	-10	= $-x_{13}$
0	1	-40	= $-x_{12}$
1	0	-30	= $-x_{21}$
-1	1*	-30	= $-x_{22}$
-3	4	380	= w

x_{13}	x_{22}	1	
1	-1	-10	$= -x_{11}$
0	1	-40	$= -x_{12}$
-1	1	-20	$= -x_{21}$
1	0	-40	$= -x_{23}$
3	1	350	$= w$

This last tableau gives us the minimal cost solution; it is

$$x_{11} = 10 \quad x_{12} = 40 \quad x_{13} = 0$$
$$x_{21} = 20 \quad x_{22} = 0 \quad x_{23} = 40$$

with a total cost of 350. (Measured in thousands, this is really \$350,000.)

Such problems, called transportation problems, can be solved by a considerably simpler method; we shall see this in a later section of this chapter.

9. DEGENERACY PROCEDURES

In case of degeneracy, i.e. assuming that some zeros appear in the right-hand column of the simplex tableau, the foregoing arguments do not hold The main difficulty is that, in Stage II for instance, the new tableau obtained need not give a better value for the objective function. But it is precisely the improvement in the objective function that guarantees that the process will terminate; in a case of degeneracy we might, with bad luck, find that the process cycles back to a previous position. Similar difficulties arise in Stage I.

From the geometric point of view, degeneracy arises when more than n of the bounding planes of the constraint set pass thorough point (n, here, is the dimension of the space). Figure II.9.1 illustrates this: it shows the set satisfying

2.9.1
$$x + z \leq 1$$
$$y + z \leq 1$$
$$x, y, z \geq 0$$

As may be seen, the planes $x + z = 1$, $y + z = 1$, $x = 0$, and $y = 0$ all pass through the point (0,0,1). This is not normal: in three-dimensional space, four planes will usually meet, three at a time, in four different points.

Precisely because this situation is not normal, we can change it by what is called a *perturbation*. Let us consider the slightly different set

FIGURE II.9.1 An example of degeneracy: four planes pass through the vertex (0,0,1).

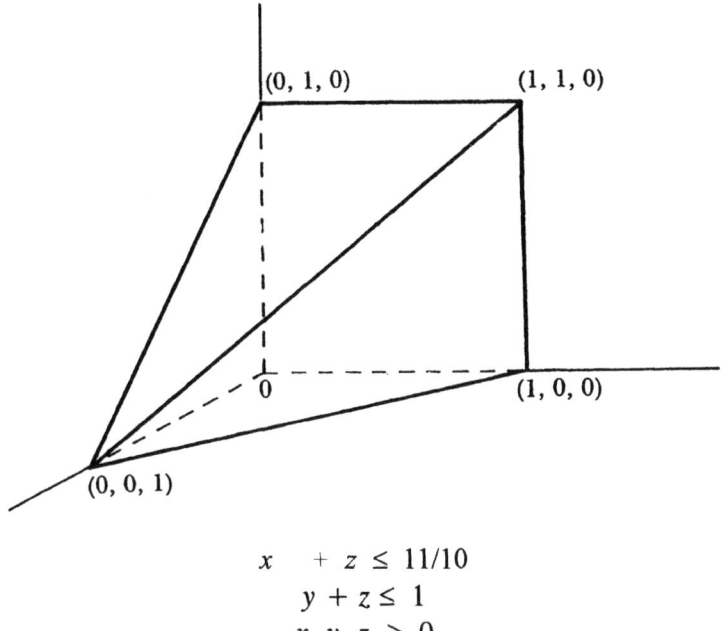

2.9.2
$$x \quad + z \leq 11/10$$
$$y + z \leq 1$$
$$x, y, z \geq 0$$

which differs from the previous system in that the first inequality has been slightly "perturbed," i.e., the right side has been changed by 1/10. Figure II.9.2 shows the new constraint set: the vertex at (0,0,1) has been split into two vertices, one of which is now at (1/10,0,1). Two of the other vertices have been slightly perturbed, the others not at all.

Suppose now that we are given a program with (2.9.1) as constraint set. We replace the constraint set by (2.9.2) and solve. The solution of the new program will be at one of the vertices of Figure II.9.2. Each of these vertices corresponds to a unique vertex of Figure II.9.1; for instance, (0,1,0) corresponds to (0,1.0), while (11/10,0,0) corresponds to (1,0,0). (The converse is not true: a vertex of Figure II.9.1 may correspond to more than one vertex of Figure II.9.2.) That vertex of Figure II.9.1 that corresponds to the solution of the program (2.9.2) will probably be the solution of the program (2.9.1); if (2.9.2) has its solution at (1/10,0,1), then (2.9.1) will probably have its solution at (0,0,1). We say "probably," because it may be that if the perturbation has been too large, the solution of the perturbed problem (2.9.2) does not correspond to the solution of (2.9.1). In that case, the perturbation should have been smaller: the first constraint may be changed, say, to $x + z \leq 101/100$.

FIGURE II.9.2 A slight perturbation of Figure II.9.1

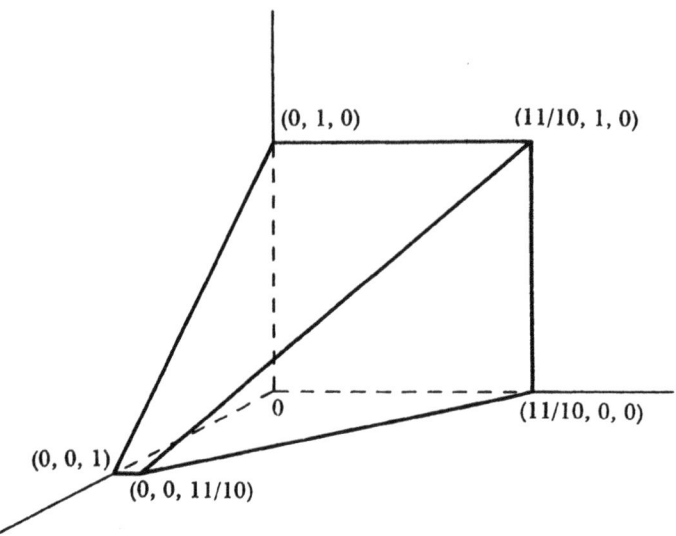

This, then, is the usual technique for perturbation. Each time a zero appears in the right-hand column, we perturb slightly, replacing the zero by a small number. Since we do not know how small is small enough to preserve the position of the solution, we generally use the letter -ε, representing a very small but negative amount. The second time that we obtain a zero in the right-hand column, we replace it by -ε^2, and so on. The solution, when obtained, will be in terms of ε. We then obtain the solution to the unperturbed problem by setting ε equal to zero.

In a sense, the technique is needlessly complicated; careful selection of the pivot will generally make it unnecessary. The procedure is described here only because it helps to complete the proof of convergence for the simplex algorithm, using rules II.5.1 and II.6.1. Normally, "cycling" occurs only when there is a very systematic choice of the pivot. A recommended procedure when using computers for the solution of linear programs is to instruct the computer to choose its pivots at random whenever the rules allow it any choice. This makes certain that the pivot is not chosen systematically, and cycling is avoided.

Although the perturbation method is not generally recommended, **we** give the following example.

II.9.1. Example

Maximize
$$w = x + y + 3z$$

Subject to
$$x + z \leq 1$$
$$y + z \leq 1$$
$$x, y, z \geq 0$$

We have here the tableau

x	y	z	1	
1	0	1	-1	$= -r$
0	1	1*	-1	$= -s$
1	1	3	0	$= w$

Let us assume we take our pivot from the third column. The two quotients are equal, and so we may pivot on either one of the entries in this column. *It is this situation that gives rise to degeneracy.* Pivoting on the starred entry, we get the tableau

x	y	s	1	
1	-1	-1	0	$= -r$
0	1	1	-1	$= -z$
1	-2	-3	3	$= w$

Note the 0 in the right-hand column. This shows degeneracy: we therefore perturb the tableau to get

x	y	s	1	
1*	-1	-1	$-\varepsilon$	$= -r$
0	1	1	-1	$= -z$
1	-2	-3	3	$= w$

Pivoting on the starred entry, we then have

r	y	s	1	
1*	-1	-1	-ε	= -x
0	1	1	-1	= -z
-1	-1	-2	3+ε	= w

Thus the solution to the perturbed problem is $x = \varepsilon$, $y = 0$, $z = 1$, with a value $w = 3 + \varepsilon$ for the objective function. We conclude that the unperturbed problem has the solution $x = 0$, $y = 0$, $z = 1$, and $w = 3$.

10. SOME PRACTICAL COMMENTS

The number of pivot steps necessary to solve an $m \times n$ program (i.e., a program with m inequalities in n variables, not counting non-negativity constraints) is generally of the order of $m+n$. This means that, in most cases, it will be between, say, $(m+n)/2$ and $2(m+n)$. This is certainly a tremendous gain over the "brute force" method described in Section II.1. For instance, if we have $m = n = 10$, we expect that the simplex algorithm will require between 10 and 40 pivot steps, whereas inspecting all the intersections of 10 of the planes would imply solving $20!/(10!\,10!) = 184{,}756$ systems of 10 equations in 10 unknowns! Nevertheless, for large grams, it is always desirable to decrease even the number of steps that the simplex method uses. This can be done by a judicious choice of pivot. In fact, rules II.3.2 and II.3.5 often give us great latitude for choice. For instance, there may be many positive entries in the bottom row. Much work has been done in perfecting the choice of pivot so as to cut the number of steps considerably, in some cases by as much as one half, over indiscriminate choices. Unfortunately, we cannot, within the scope of this book, study all these systems. Interested readers should read more advanced treatises on linear programming.

PROBLEMS ON THE SIMPLEX ALGORITHM

1. Maximize

$$w = 2x + 3y + z$$

Subject to

$$x + 2y - 4z \leq 10$$
$$3y + z \leq 7$$
$$2x - 3y + 4z \leq 12$$
$$x, y, z \geq 0$$

2. Minimize
$$w = x + 2y + z$$

Subject to
$$3x + y + 2z \geq 20$$
$$x - 3y + 4z \geq 15$$
$$-x + y + z \geq 8$$
$$x, y, z \geq 0$$

3. Maximize
$$w = x - 2y + 4z$$

Subject to
$$x + y + 2z \leq 20$$
$$x - y + z \leq 5$$
$$2x + 3y + 2z \leq 35$$
$$x, y, z \geq 0$$

4. Minimize
$$w = 5x + 2y + 3z$$

Subject to
$$3x - 2y + 6z \geq 25$$
$$x + y + 2z \geq 15$$
$$3x + y - 3z \geq 10$$
$$x, y, z \geq 0$$

5. A dietitian is given three foods. Each unit of food A contains 10 oz. protein, 4 oz. fat, and 6 oz. carbohydrate, and costs \$1. A unit of food B has 4 oz. protein, 6 oz. fat, and 8 oz. carbohydrate, and costs 50¢. A unit of food C contains 2 oz. protein, 12 oz. fat, and 8 oz. carbohydrate, and costs 30¢. A mixture of these three foods must be made, containing at least 50 oz. protein, 60 oz. fat, and 65 oz. carbohydrate. What is the cheapest such mixture possible?

6. A toy company makes three types of toy, which must be processed through three machines called a twister, a bender, and a painter. Toy A requires 1 hr. in the twister, 2 hr. in the bender, and 1 hr. in the painter, and sells for $4. Toy B requires 2 hr. in the twister, 1 hr. in the bender, and 3 hr. in the painter, and sells for $5. Toy C requires 3 hr. in the twister, 2 hr. in the bender, and 1 hr. in the painter, and sells for $9. The twister can work at most 50 hr. per week; the bender, at most 40 hr.; and the painter, at most 60 hr. How many toys of each type should the company produce to maximize revenues?

7. A furniture firm manufactures chairs, tables, and beds, and has three workers. The time in hours that each employee must spend to produce one of these, and the corresponding profit, are given in the table:

	A	B	C	Profit ($)
Chair	1	3	2	10
Table	3	5	4	25
Bed	5	4	8	35

Employee A can work only 20 hr. per week, while B and C can work 50 hr. each. How many chairs, beds, and tables should be made so as to maximize profits?

8. A food processing company has 10,000 lb. of African coffee, 12,000 lb. of Brazilian coffee, and 7000 lb. of Colombian coffee. It sells four blends of coffee, whose contents (in ounces of the coffee per pound of blend) and prices are given in the table.

	African	Brazilian	Colombian	Price (¢)
A	0	0	16	90
B	0	12	4	70
C	6	8	2	60
D	10	6	0	50

How many pounds of each blend should be produced to maximize revenues?

9. A candy manufacturer has 500 lb. of chocolate, 100 lb. of nuts, and 50 lb. of fruit in inventory. He produces three types of candy. A box of type A uses 3 lb. chocolate, 1 lb. nuts, and 1 lb. fruit, and sells for $10. A box of type B uses 4 lb.

chocolate and 1/2 lb. nuts, and sells for $6. A box of type C contains 5 lb. chocolate, and sells for $4. How many boxes of each type should be made to maximize revenues?

10. A metal company needs 50 tons of iron, 5 tons of copper and 1 ton of nickel to fulfill a contract. It is offered three types of ore. Type A contains 10 per cent iron and 1 per cent copper, and costs $5 per ton. Type B contains 5 per cent copper and 2 per cent nickel, and sells for $15 per ton. Type C contains 3 per cent iron, 1 per cent copper and 2 per cent nickel, and sells for $6 per ton. How many tons of each ore should be bought to minimize costs?

11. DUALITY

Let us consider the two linear programs

2.11.1 Maximize
$$w = c_1x_1 + c_2x_2 + \ldots + c_nx_n$$

Subject to

2.11.2
$$a_{11}x_1 + a_{12}x_2 + \ldots + a_{1n}x_n \le b_1$$
$$a_{21}x_1 + a_{22}x_2 + \ldots + a_{2n}x_n \le b_2$$
$$\ldots$$
$$a_{m1}x_1 + a_{m2}x_2 + \ldots + a_{mn}x_n \le b_m$$

2.11.3
$$x_1, x_2, \ldots x_n \ge 0$$

and

2.11.4 Minimize
$$z = b_1y_1 + b_2y_2 + \ldots + b_my_m$$

Subject to

2.11.5
$$a_{11}y_1 + a_{21}y_2 + \ldots + a_{m1}y_m \ge c_1$$
$$a_{12}y_1 + a_{22}y_2 + \ldots + a_{m2}y_m \ge c_2$$
$$\ldots$$
$$a_{1n}y_1 + a_{2n}y_2 + \ldots + a_{mn}y_m \ge c_n$$

2.11.6
$$y_1, y_2, \ldots y_m \ge 0$$

In matrix notation, these programs are

2.11.7 Maximize
$$w = c'x$$

Subject to

2.11.8 $Ax \leq b$
2.11.9 $x \geq 0$

and

2.11.10   Minimize
$$z = b'y$$
Subject to

2.11.11 $A'y \geq c$
2.11.12 $y \geq 0$

where the matrix A and the vectors b and c are the same for both programs. We shall call the program (2.11.10) to (2.11.12) - or (2.11.4) to (2.11.6) - the *derived program* of program (2.11.7) to (2.11.9) - or (2.11.1) to (2.11.3).

Let us see whether we can find the derived program of (2.11.10) to (2.11.12). We notice, of course, that (2.11.10) to (2.11.12) is not the same type of program as (2.11.7) to (2.11.9), since it is a minimization problem, and, moreover, the inequalities (2.11.11) go the wrong way. We dispose of these difficulties by multiplying the objective function and the constraints (2.11.11) by -1; this gives us a restatement of (2.11.10) to (2.11.12) in the form

Maximize
$$-z = (-b')y$$
Subject to
$$(-A')y \leq -c$$
$$y \geq 0$$

We can now form the derived program of this, it is done by taking the transpose of the coefficient matrix and interchanging the two vectors -b and -c. We then obtain

2.11.13   Minimize
$$r = (-c)'s$$
Subject to
2.11.14 $(-A^t)'s \geq -b$
2.11.15 $s \geq 0$

We can, however, do the same to this program as we just did to (2.11.10) to (2.11.12). Using the fact that $A^{tt} = A$, and multiplying through by -1, we have the equivalent program

2.11.16   Maximize
$$-r = c's$$
Subject to
2.11.17 $As \leq b$
2.11.18 $s \geq 0$

It may be seen that this program (2.11.16) to (2.11.18) is the same as (2.11.7) to (2.11.9): in fact, the only difference between the two lies in fact that the variable vector is x in one case, and s in the other. In other words, the only difference lies in the labeling of the variables. This is no difference at all. We find thus, that the derived of the derived program (2.11.10) to (2.11.12) is the original program (2.11.7) to (2.11.9). When a relation is of this sort (in which each of two systems may be obtained from the other by the same set of rules), it is generally called a *duality* relation. The two programs, in this case, are said to be *dual* to each other.

II.11.1. Definition. The two programs (2.11.1) to (2.11.3) and (2.11.4) to (2.11.6) are *dual* programs. Each of these two programs is the *dual* of the other.

From a purely theoretical point of view, each of the two programs (2.11.1) to (2.11.3) and (2.11.4) to (2.11.6) is the dual of the other; hence neither enjoys any primacy over the other. From the practical point of view, of course, this is not so. The decision-maker is given a practical problem, which he expresses (if possible) in the form of a linear program. The dual program has, in general, no practical interpretation (though we shall see that it may be given one in terms of so-called *shadow prices*). Thus, the *original* practical program does enjoy a definite primacy and will, therefore, be called the *primal*. The other problem will be called the *dual*.

The relation between two dual linear programs is extremely strong, as is evidenced by the theorems that follow.

II.11.2. Theorem. Let the vectors $\mathbf{x} = (x_1, x_2, \ldots, x_n)$ and $\mathbf{y} = (y_1, \ldots, y_m)$ satisfy the constraints (2.11.8) and (2.11.9), and (2.11.11) and (2.11.12), respectively. Then
$$\mathbf{c}'\mathbf{x} \leq \mathbf{b}'\mathbf{y}$$

Proof. We prove this theorem by the following sequence of relations:
By the commutative law for vector multiplication,
$$\mathbf{x}'\mathbf{c} = \mathbf{c}'\mathbf{x}$$

Now, $\mathbf{c} \leq A'\mathbf{y}$, and $\mathbf{x} \geq 0$, and so
$$\mathbf{x}'\mathbf{c} \leq \mathbf{x}'(A'\mathbf{y}).$$

By the associative law for matrix multiplication,
$$\mathbf{x}'(A'\mathbf{y}) = (\mathbf{x}'A')\mathbf{y}$$

Now $A\mathbf{x} \leq \mathbf{b}$ so $\mathbf{x}'A' \leq \mathbf{b}'$. Since $\mathbf{y} \geq 0$, then
$$(\mathbf{x}'A')\mathbf{y} \leq \mathbf{b}'\mathbf{y}$$

This gives us the chain of relations
$$\mathbf{c}'\mathbf{x} = \mathbf{x}'\mathbf{c} \leq \mathbf{x}'(A'\mathbf{y}) = (\mathbf{x}'A')\mathbf{y} \leq \mathbf{b}'\mathbf{y}$$

which proves the theorem.

II.11.3. Corollary. Suppose $\mathbf{x}^*$ and $\mathbf{y}^*$ satisfy the constraints (2.11.8) to (2.11.9) and (2.11.11) to (2.11.12) respectively) and, moreover,
$$\mathbf{c}'\mathbf{x}^* = \mathbf{b}'\mathbf{y}^*$$

Then $\mathbf{x}^*$ and $\mathbf{y}^*$ are the solutions of the programs (2.11.7) to (2.11.9) and (2.11.10) to (2.11.12) respectively.

Proof. This is clear since, if $\mathbf{x}$ is any vector satisfying (2.11.8) to (2.11.9), then, by Theorem II.11.2,

$$c'x \leq b'y^* = c'x^*$$

Hence **x*** solves (2.11.7) to (2.11.9), and similarly, **y*** solves (2.11.10) to (2.11.12).

Corollary II.11.3 gives us a method of checking whether a given pair of vectors **x*** and **y*** solve the two dual programs (2.11.7) to (2.11.9) and (2.11.10) to (2.11.12), respectively. In fact, if they satisfy the constraints of the Programs, and give the same values for the objective functions, then II.11.3 confirms that they are solutions. The converse of II.11.3 is also true; if **x*** and **y*** are solutions to the two mutually dual programs (2.11.7) to (2.11.9) and (2.11.10) to (2.11.12) then $c'x^* = b'y^*$. Thus equality of values is both necessary and sufficient for the vectors **x*** and **y*** to solve the two programs. (We shall prove this statement later.) We give an example of this property now.

II.11.4. Example. Consider the two programs

Maximize
$$w = 3x_1 + 2x_2 + 4x_3$$

Subject to
$$x_1 + 3x_2 + 2x_3 \leq 10$$
$$2x_1 + x_2 + x_3 \leq 8$$
$$x_1, x_2, x_3 \geq 0$$

and
Minimize
$$z = 10y_1 + 8y_2$$

Subject to
$$y_1 + 2y_2 \geq 3$$
$$3y_1 + y_2 \geq 2$$
$$2y_1 + y_2 \geq 4$$
$$y_1, y_2 \geq 0$$

The maximizing problem gives us the tableau

	x_1	x_2	x_3	1	
	1	3	2*	-10	$= -u_1$
	2	1	1	8	$= -u_2$
	3	2	4	0	$= w$

Pivoting on the starred entries, we obtain the tableaux

	x_1	x_2	u_1	1	
	1/2	3/2	1/2	-5	$= -x_3$
	3/2*	-1/2	-1/2	-3	$= -u_2$
	1	-4	-2	20	$= w$

and

	u_2	x_2	u_1	1	
	1/3	2/3	5/3	-4	$= -x_3$
	2/3	-1/3	-1/3	-2	$= -x_1$
	-2/3	-11/3	-5/3	22	$= w$

which gives us the solution; it is $x_1 = 2$, $x_2 = 0$, $x_3 = 4$, and gives a value $w=22$. Consider, now, the minimizing problem. Its tableau is

	y_1	y_2	1	
	-1	-2	3	$= -v_1$
	-3	-1	2	$= -v_2$
	-2*	-1	4	$= -v_3$
	10	8	0	$= z$

We pivot on the starred entries to obtain

	v_3	y_2	1	
	-1/2	-3/2*	1	$= -v_1$
	-3/2	1/2	-4	$= -v_2$
	-1/2	1/2	-2	$= -y_1$
	5	3	20	$= z$

and

	v_3	v_1	1	
	1/3	-2/3	-2/3	$= -y_2$
	-5/3	1/3	-11/3	$= -v_2$
	-2/3	1/3	-5/3	$= -y_1$
	4	2	22	$= z$

which gives the solution; $y_1 = 5/3$, $y_2 = 2/3$ with a value $z = 22$. It is now possible to check that the given vectors are, indeed, the solutions to the two programs. In fact, they satisfy the constraints, and give the same value for the objective functions.

Looking over the solutions of the pair of dual programs, we can see that the implications of duality are far greater than the simple fact that the values of the two programs (i.e., the maximum and minimum values, respectively, of their objective functions) are equal. In fact, it is not only the objective functions that give the same value: the corresponding tableaux for the two programs are, at each step, dually related. Each tableau is "almost" the negative transpose of the corresponding tableau for the other problem. That this is so for any pair of dual programs is immensely important.

Let us, by the introduction of slack variables, rewrite (2.11.7) to (2.11.9) in the form

2.11.19   Maximize
$$w = c'x$$
Subject to
2.11.20 $Ax - b = -u$
2.11.21 $x, u \geq 0$

Similarly, we may rewrite (2.11.10) to (2.11.12) in the form

2.11.22   Minimize
$$z = b'y$$

Subject to

2.11.23 $A'y - c = v$
2.11.24 $y, v \geq 0$

We may write these two programs simultaneously in the tableau

2.11.25

	x_1	x_2	....	x_n	1	
y_1	a_{11}	a_{12}	....	a_{1n}	$-b_1$	$= -u_1$
y_2	a_{21}	a_{22}	....	a_{2n}	$-b_2$	$= -u_2$
....	....	....	....	....	....	....
....	....	....	....	....	....	....
....	....	....	....	....	....	....
y_m	a_{m1}	a_{m2}	....	a_{mn}	$-b_n$	$= -u_m$
-1	c_1	c_2	....	c_n	0	$= w$
	$= v_1$	$= v_2$	....	$= v_n$	$= -z$	

The rows of (2.11.25) represent, as usual, the equations (2.11.19) and 2.11.20); the columns, now, represent the equations (2.11.22) and (2.11.23). Thus, the single tableau does double duty as a representation of the pair of problems simultaneously. The pivot steps of the simplex algorithm were devised precisely to maintain the validity of the row equations. What we will show is that they have the property of preserving the column equations as well. Let us, then, consider the column system of a tableau.

2.11.26

	a_{11}	a_{12}	....	a_{1n}	$-b_1$
q_1					
q_2	a_{21}	a_{22}	....	a_{2n}	$-b_2$
....	....	....	....	....	....
....	....	....	....	....	....
....	....	....	....	....	....
q_m	a_{m1}	a_{m2}	....	a_{mn}	$-b_n$
-1	c_1	c_2	....	c_n	δ
	$= p_1$	$= p_2$	....	$= p_n$	$= -z$

In (2.11.26), the p_j are the basic variables, while the q_i are non-basic. Suppose we wish to interchange p_j and q_i. The jth column represents the equation

2.11.27
$$a_{1j}q_1 + a_{2j}q_2 + ... + a_{ij}q_i + ... - c_j = p_j$$

We can solve (2.11.27) for q_i, obtaining

2.11.28
$$(a_{1j}/a_{ij})q_1 - (a_{2j}/a_{ij})q_2 + ... + (1/a_{ij})p_j + ... + (c_j/a_{ij}) = q_i$$

In turn, the lth column represents

2.11.29 $\qquad a_1q_1 + a_2q_2 + \ldots + a_iq_i + \ldots - c_l = p_l$

Let us, now, substitute (2.11.28) in (2.11.29). This gives us

2.11.30 $(a_{1l} - a_{il}a_{1j}/a_{ij})q_1 + (a_{2l} - a_{il}a_{2j}/a_{ij})q_2 + \ldots + (a_{il}/a_{ij})p_j + \ldots + (c_j - a_{il}c_j/a_{ij}) = p_l$

The changes in the coefficients can be represented by the schema

2.11.31

a	b
c	d

→

$1/a$	b/a
$-c/a$	$d - bc/a$

in which, as in Section II.2, a represents the pivot, b is an entry in the pivot row, c is an entry in the pivot column, and d the entry in the row of c and column of b. But (2.11.31) is exactly the same as (2.3.10). Thus the pivot transformations that preserve the row equations also preserve the column equations.

Let us suppose that the program (2.11.1) to (2.11.4) has a solution. This solution is obtained, as we know, from a tableau in which all entries in the right-hand column and in the bottom row, (except possibly the bottom right-hand entry) are non-positive. Consider, then, the column equations in this tableau:

2.11.32

q_1		a_{11}	a_{12}	....	a_{1n}	$-b_1$
q_2		a_{21}	a_{22}	....	a_{2n}	$-b_2$
....		....	....	....	....	....
....		....	....	....	....	....
....		....	....	....	....	....
q_m		a_{m1}	a_{m2}	....	a_{mn}	$-b_n$
-1		c_1	c_2	....	c_n	δ
		$= p_1$	$= p_2$	....	$= p_n$	$= -z$

It may be seen from (2.11.32) that, if we set the non-basic variables $q_1, q_2, \ldots, q_m$ equal to zero, each of the basic variables, will be equal to the negative of the corresponding entry in the bottom row: $p_j = -c_j$. But, by hypothesis, $c_j \leq 0$. Hence $p_j \geq 0$. Thus the tableau (2.11.32) gives a b.f.p. for the dual program (2.11.4) to (2.11.6).

Consider, next, the right-hand column of (2.11.32). It represents the equation

2.11.33 $\qquad z = b_1q_1 + b_2q_2 + \ldots + b_mq_m + \delta$

Now, by hypothesis, each of the terms $b_j q_j$ is non-negative (being the product of a non-negative constant b_j and a non-negative variable q_j). It follows that, for every point in the constraint set, $z \geq \delta$. We have however, a point that gives a value $z = \delta$ for the objective function.

It follows that the b.f.p. of (2.11.32) is the solution of (2.11.4) to (2.11.6); the solutions to both programs (2.11.1) to (2.11.3) and (2.11.4) to (2.11.6) are given by the same tableau (2.11.32). Therefore, optimal values of the objective functions must be the same for both programs, i.e., δ. We obtain the following theorem, known as the fundamental theorem of linear programming.

II.11.5 Theorem. If either one of the two programs (2.11.1) to (2.11.3) and (2.11.4) to (2.11.6) has a solution then so does the other, and the solutions give the same value to the objective functions. If both programs are feasible, then both have solutions. If one of the two programs is feasible but unbounded, then the other is infeasible. Finally, if one of the two programs is infeasible, then the other is either infeasible or unbounded.

Proof: The first statement of this theorem has just been proved. The other statements depend on the fact that a program has a solution if it is bounded and feasible (seen in our development of the simplex algorithm), as well as on Theorem II.11.2.

That the same tableau solves both of a pair of dual problems is additionally advantageous because very often the first tableau obtained is a b.f.p., not for the *primal* problem, but for the *dual*. By attacking the dual, rather than the primal, we are able to go into Stage II immediately, avoiding perhaps half of the pivot steps normally necessary.

II.11.6. Theorem (Complementary Slackness).
Let **x** and **y** be vectors satisfying the constraints of programs (2.11.7) to (2.11.9) and (2.11.10) to (2.11.12) respectively. Then a necessary and sufficient condition for **x** and **y** to be the solutions of the two programs is that, for every j,

$$\underline{\text{either}}\ \ x_j = 0 \ \ \underline{\text{or}}\ \ \Sigma a_{ij} y_i = c_j$$

and that, for every i,

$$\underline{\text{either}}\ \ y_i = 0 \ \ \underline{\text{or}}\ \ \Sigma a_{ij} x_j = b_i.$$

Proof. We have already seen that a necessary and sufficient condition is that b'y = x'c

Now, introducing slack variables, as in (2.11.19) to (2.11.21) and (2.11.22) to (2.11.24) respectively, we have

$$b'y - x'c = b'y - (x'A')y + x'(A'y) - x'c$$
$$= (b'-x'A') y + x'(A'y-c)$$
$$= u'y - x'v.$$

Since **u, y, x,** and **v** are all non-negative, it follows that this last expression can be zero only if, for each j, either x_j or v_j is 0, and, for each i, either y_i or u_i is zero. But this condition is also sufficient.

This last theorem is valuable in that it allows us to check for optimality without the need for evaluating the objective functions.

II.11.7. Example. A mixture is to be made of foods, A, B, and C. Each unit of food A contains 3 oz. protein and 4 oz. carbohydrate, and costs \$1. Each unit of food B contains 5 oz. protein and 3 oz. carbohydrate, and costs \$3. Each unit of food C contains 1 oz. protein and 4 oz. carbohydrate, and costs 50'. The mixture is to contain at least 15 oz. protein and 10 oz. carbohydrate. Find the mixture of minimal cost.

The program here is:

Minimize
$$z = y_1 + 3y_2 + 1/2\, y_3$$
Subject to
$$3y_1 + 5y_2 + y_3 \geq 15$$
$$4y_1 + 3y_2 + 4y_3 \geq 10$$
$$y_1, y_2, y_3 \geq 0$$

The first tableau obtained for this program will not be a b.f.p.: **y** = **0** is not a feasible vector. On the other hand, we may write the tableau that represents this program as the column system: the dual will then be the row system:

	x_1	x_2	1	
y_1	3*	4	-1	$= -u_1$
y_2	5	3	-3	$= -u_2$
y_3	1	4	-1/2	$= -u_3$
-1	15	10	0	$= w$
	$= v_1$	$= v_2$	$= -z$	

It may be seen that this is a b.f.p. for the dual problem. We pivot on the starred element, obtaining

	u_1	x_2	1	
v_1	1/3	4/3	-1/3	$= -x_1$
y_2	-5/3	-11/3	-4/3	$= -u_2$
y_3	-1/3	8/3	-1/6	$= -u_3$
-1	-5	-10	5	$= w$
	$= y_1$	$= v_2$	$= -z$	

which gives the solution; $y_1 = 5$, $y_2 = 0$, $y_3 = 0$ with a value $z = 5$. Note how Stage I was avoided by solving the dual program.

In a pair of mutually dual programs, generally one of the pair has a clear practical application: it has been formulated, precisely, to solve a practical problem. Yet, what about the dual? Is it possible to obtain a "practical" interpretation of the dual program?

Let us consider the dual of Example II.11.7.

Maximize
$$15x_1 + 10x_2$$

Subject to
$$3x_1 + 4x_2 \leq 1$$
$$5x_1 + 3x_2 \leq 3$$
$$x_1 + 4x_2 \leq 1/2$$
$$x_1, x_2 \geq 0$$

and see whether some interpretation may be obtained. Each of the two variables, x_1 and x_2, corresponds to one of the constraints. But these constraints correspond to the two nutrients, protein and carbohydrate. Thus, x_1 and x_2 are numbers assigned, respectively, to protein and carbohydrate; the type of number can be seen from the fact that the constants in the constraints correspond to prices - the prices of the different foods, A, B, and C. It follows that we may consider x_1 and x_2 as representing some sort of value, in dollars per ounce, for the two nutrients. We are assigning such values, known as *shadow prices*, to these nutrients. The constraints represent, in a sense, a profitless system: each of three foods costs at least as much as the values of the nutrients it contains. The program seeks to maximize the total shadow prices of the required diet.

The solution of the dual problem is given by the tableau: it is $x_1 = 1/3$, $x_2 = 0$, with a value $w = 5$. Note that this seems to make the carbohydrate valueless. This is not to say that the carbohydrate is useless; rather, it represents the well-known economic fact that surplus goods lose all commercial value, and in this case the solution to the primal problem gives a surplus of carbohydrate. Note also that the solution of the primal problem does not use any of foods B and C, coincident with the fact that the costs of these foods are higher than the shadow prices of the nutrients contained (according to the dual solution).

Generally speaking, the dual may be interpreted in terms of shadow prices whenever the primal problem deals with costs or profits.

II.11.8. Example. A company makes three products, A, B, and C, which must be processed by three employees, D, E, and F. Each unit of product A must be processed 1 hr. by employee D, 2 hr. by E, and 3 hr. by F, and yields a profit of $50. Each unit of B must be processed 4 hr. by D, 1 hr. by E and 1 hr. by F, and yields a profit of $30. Each unit of C must be processed 1 hr. by D, 3 hr. by E, and 2 hr. by F, and yields a profit of $20. Employee D can work at most 40 hr. per week, E, at most 50 hr., and F, at most 30 hr. It is required to maximize the profit.
The program here is
   Maximize
$$w = 50x_1 + 30x_2 + 20x_3$$
Subject to
$$x_1 + 4x_2 + x_3 \leq 40$$
$$2x_1 + x_2 + 3x_3 \leq 50$$
$$3x_1 + x_2 + 2x_3 \leq 30$$
$$x_1, x_2, x_3 \geq 0$$

The dual program is:

Minimize
$$z = 40y_1 + 50y_2 + 30y_3$$

Subject to
$$y_1 + 2y_2 + 3y_3 \geq 50$$
$$4y_1 + y_2 + y_3 \geq 30$$
$$y_1 + 2y_2 + 2y_3 \geq 20$$
$$y_1, y_2, y_3 \geq 0$$

which may be interpreted as the assignation of natural wages to the three employees. It may be checked that the solutions to the two problems are

$$x_1 = 80/11, x_2 = 90/11, x_3 = 0, w = 6700/11$$

for the primal, and

$$y_1 = 40/11, y_2 = 0, y_3 = 170/11, z = 6700/11$$

for the dual. Note that at the optimum, $y_2 = 0$. This should not be taken to mean that employee E should receive no wages; rather, he should be encouraged to work a shorter week, since he is idle more than half the time. Nor does our solution suggest that F should be paid more than $15 per hour; it does say that his week should be lengthened if extra hours for him will cost less than $15.45 per hour. The shadow wages tend to represent the marginal productivity of the three employees concerned. (A classical economist would say that F's wages would tend to seek the higher level causing him to work longer hours until an equilibrium was found.)

PROBLEMS ON DUALITY

1. Give the duals of the following linear programs. Obtain solutions to each pair of programs:

(a) Minimize
$$2x + 4y + 7z = w$$

Subject to
$$3x + y + 2z \leq 25$$
$$x + 3y - z \leq 10$$
$$4y + z \leq 30$$
$$x, y, z \geq 0$$

(b) Minimize
$$6x + 2y + 3z = w$$

Subject to
$$x + 4y - 3z \geq 15$$
$$2x + 3y + z \geq 25$$
$$-x + 4y + 2z \geq 12$$
$$x, y, z \geq 0$$

(c) Minimize
$$3x + 2y + z =$$

Subject to
$$x + y + 3z \leq 25$$
$$x - y \leq -10$$
$$y + 2z \leq 15$$
$$x, y, z \geq 0$$

(d) Maximize
$$3x + 2y + z = w$$

Subject to
$$x + 3y + z \geq 30$$
$$x - y > -10$$
$$3x + y + z \geq 40$$
$$x, y, z \geq 0$$

2. Give the duals of problems 5 to 10 on pages 75-77 . Solve these duals and give a heuristic interpretation (shadow prices).

3. Show that, to an equation constraint in the primal problem, there corresponds an unrestricted (i.e., not necessarily non-negative) variable in the dual, and vice-versa.

12. TRANSPORTATION PROBLEMS

The simplex algorithm is probably the most efficient method available for the solution of linear programs in general. This does not mean, however, that it will be uniformly the most efficient method. The reason is that there are often special types of problem that, precisely because of their special type, can be handled more efficiently by a method other than the simplex algorithm. Much of the present research in linear programming consists precisely in the search for new methods of solving special types of linear programs - generally those that the researcher (an applied mathematician or operations analyst) has encountered in practical experience. Some of these new methods are tremendously efficient - for the particular type of problem concerned, generally considerably larger than could be solved with the simplex algorithm.

We consider here a special type of linear program: the *network* program. Such a program was seen earlier as Example II.8.1. Its characteristic property is that, as may be seen from inspection of the several tableaux in II.8.1, the entries in the inner part of the tableau (i.e., other than on the bottom row or right-hand column) are always 0, 1, or -1. The importance of this fact is that the pivot is always 1 or -1; in effect, this means that no multiplications or divisions will ever be necessary - the simpler operations of addition and subtraction will always be sufficient for solution.

Example II.8.1 is a typical *transportation problem*. In the general case, we might picture a company with a total of m warehouses, each containing a certain amount of the company's product, and n distributors, each having a demand for a certain amount of the product. The problem is to transport the desired amounts from the warehouses to the distributors in such a way as to minimize shipping costs.

More precisely, let a_i be the availability at the ith warehouse; let b_j be the requirement at the jth distributor. We make the assumption:

2.12.1
$$\sum_{i=1}^{m} a_i = \sum_{j=1}^{n} b_j$$

which means that the total availabilities at the m warehouses are exactly equal to the total requirements at the n distributors. Let c_{ij} be the cost of transporting a unit from the ith warehouse to the jth distributor. The problem can then be expressed as a linear program:

2.12.2 Minimize

$$\sum_{i=1}^{m}\sum_{j=1}^{n} c_{ij} x_{ij}$$

Subject to

2.12.3 $\quad\sum_{i=1}^{m} x_{ij} = b_j \quad$ for each j

2.12.4 $\quad\sum_{j=1}^{n} x_{ij} = a_i \quad$ for each i

2.12.5 $\quad x_{ij} \geq 0 \quad$ for each i, j

The first thing to notice is that there is no need for slack variables, since all the constraints (2.12.3) and (2.12.4) are equations rather than inequalities. This means that the "flows" x_{ij} are the only variables. Since each b.f.p. of a program is determined by the non-basic variables, to determine a b.f.p. of this program, it is sufficient to state which of the variables x_{ij} are equal to zero. This being so, it follows that we can represent each b.f.p. by a schematic graph.

Technically speaking, a graph is a collection of points called *nodes* and segments of curves called *arcs* that connect some of the nodes. (This is a *topological* graph, as distinguished from the graph of a function.) To each b.f.p. we assign a graph as follows: there is a node W_i for each of the warehouses and a node D_j for each of the distributors. There will be an arc joining W_i to D_j if x_{ij} is positive, and there will be no other arcs. Thus, the tableau for the first b.f.p. reached in the solution of Example II.8.1 can be represented either by a graph or by a matrix:

	D_1	D_2	D_3
W_1	0	10	40
W_2	30	30	0

We repeat that the graph determines the b.f.p. entirely; in fact, the "missing" arcs, i.e., from W_1 to D_1 and from W_2 to D_3 in this example, correspond to the non-basic variables. These determine the values of the basic variables entirely.

The following theorem is a fundamental point of the method of solution of transportation problems. We say a graph is connected if any two of the nodes can be joined by a sequence of arcs from the graph.

II.12.1. Theorem. The graph corresponding to a b.f.p. will not contain any closed loops. Moreover, assuming non-degeneracy of the program, the graph will be connected, and so the addition of any arc will close a loop.

Proof. Let us suppose that the graph corresponding to a b.f.p. does contain a closed loop. This loop must have an even number of arcs, since each arc connects a warehouse to a distributor. Let α be the smallest flow along any of the arcs in the loop. If we number these arcs consecutively, it may be seen that we can increase the flow along the odd-numbered arcs by $\alpha/2$ and decrease it along the even-numbered arcs by the same amount without contradicting any of the constraints. This gives us a different point that has the same arcs "missing" from its graph, which contradicts the fact that the non-basic variables entirely determine a b.f.p. The contradiction proves the first statement of the theorem.

To prove the second statement, we point out that the graph can be disconnected only if the availabilities at a subcollection of the warehouses are equal to the requirements at a subcollection of the distributors. This is, precisely, degeneracy. Assuming that the graph is connected, any new arc will close a loop, since it will connect two nodes that were already connected in some other manner.

Now that we have replaced the simplex tableau by a graph we must see what form the pivot-step takes. We know that a pivot-step has the property of increasing the value of one of the non-basic variables, and it will, thus, correspond to adding one of missing arcs to the graph. This will cause one of the arcs presently on the graph to be removed. Let us see which one.

In the preceding graph, there are two **missing arcs:** W_1D_1 and W_2D_3 (corresponding to $x_{11} = x_{23} = 0$). We can make x_{11} positive by adding the arc W_1D_1 to the graph: this closes a loop, which we denote as $W_1D_1W_2D_2W_1$. Suppose that we do increase the variable x_{11}. This means that because a smaller quantity of the product is left at W_1, we must decrease either x_{12} or x_{13} (the flows from W_1 to the other two D_j).

We cannot decrease x_{13}, as this would entail an increase in x_{23}, and we want to keep x_{23} fixed at zero (the fundamental idea of pivot step). Hence we decrease x_{12}. Now the requirement at D_2 is not being met, and this deficiency must be made up by increasing x_{22}. In turn, this causes x_{21} to decrease. The decrease in x_{21} will be exactly enough to meet the increase in x_{11}, keeping the required quantity at D_1. Thus, an increase in x_{11} will cause an equal increase in x_{22} and corresponding equal decreases in x_{12} and x_{21}. We increase x_{11}, then, until either x_{12} or x_{21} has decreased to zero. Since $x_{12} = 10$ and $x_{21} = 30$, It is clear that x_{12} will be the first to vanish; adding the arc $W_1 D_1$ to the graph has caused us to remove the arc $W_1 D_2$. We obtain the new b.f.p.

	D_1	D_2	D_3
W_1	10	0	40
W_2	20	40	0

It remains to be seen whether this new b.f.p. is an improvement over the previous one. This can actually be calculated directly. If we do, we see that the total costs for this b.f.p. are 510, whereas, for the previous b.f.p., they were only 500. Since we wish to minimize costs, this is not an improvement. On the other hand, we could also have seen that the change would increase costs in the following manner: each unit increase in x_{11} causes an equal increase in x_{22}, and equal decreases in x_{21} and x_{12}. The total change in costs for a unit change in x_{11} is therefore

$$c_{11} - c_{12} + c_{22} - c_{21} = 1 - 4 + 9 - 5 = 1$$

Since this is positive, we would conclude that the change would cause an increase in total costs, and therefore reject it.

This describes, then, the nature or the fundamental step for the transportation problem. Under the assumption of non-degeneracy, Theorem II.12.1 tells us that each missing arc will, when added to the graph, close a loop. If we number the arcs in this loop consecutively, starting with the newly included arc, we see that an increase along the first arc will cause equal increases along the odd numbered arcs, and decreases along the even-numbered arcs. To see whether she change will be an improvement, we add the unit costs along the odd arcs, and subtract from these the unit costs along the even arcs. If the result is negative the change is an improvement (it will decrease costs). The new arc is then added to the diagram, while one of the even arcs in the loop (the one with the lowest flow) is removed.

With this method explained, we proceed to solve the problem of Example II.8.1.

II.12.2. Example. Solve the 2 × 3 transportation problem with availabilities

$$a_1 = 50 \qquad\qquad a_2 = 60$$

and requirements

$$b_1 = 30 \qquad b_1 = 30 \qquad b_1 = 30$$

and transportation costs given by the table:

TABLE II.12.1
From Warehouse To Distributor

	D_1	D_2	D_3
W_1	1	4	1
W_2	5	9	2

We already have obtained a b.f.p., given by

	D_1	D_2	D_3
W_1	0	10	40
W_2	30	30	0

Of the two missing arcs, we have already seen that W_1D_1 will not cause a favorable change. Consider, then, the arc W_2D_3. This will close the loop $W_2D_3\,W_1D_2W_2$. The change in cost per unit is given by

$$c_{23} - c_{13} + c_{12} - c_{22} = 2 - 1 + 4 - 9 = -4$$

Since this is negative, the change is a favorable one. We therefore add the arc W_2D_3 to the graph. Of the even arcs in the loop, W_1D_3 and W_2D_2, we see that W_2D_2 has the smaller flow: $x_{13} = 40$ and $x_{22} = 30$. Hence, we remove W_2D_2 from the graph; x_{23} and x_{12} increase by 30, while x_{13} and x_{22} decrease by 30.

	D_1	D_2	D_3
W_1	0	40	10
W_2	30	0	30

We repeat the procedure. Of the two arcs missing in the new graph, W_2D_2 will clearly not give any improvement; we would simply be replacing the arc we removed in the previous step. Consider, then, the arc W_1D_1. This closes the loop $W_1D_1W_2D_3W_1$. We calculate

$$c_{11} - c_{21} + c_{23} - c_{13} = 1 - 5 + 2 - 1 = -3$$

This also is negative, so W_1D_1 is added to the graph. We see, next, that $x_{21} = 30$, while $x_{13} = 10$. The arc W_1D_3 should be removed. We increase x_{11} and x_{23} by 10 while decreasing x_{21} and x_{13} by the same amount. This gives us

	D_1	D_2	D_3
W_1	10	40	0
W_2	20	0	40

This new schema is the solution. Of the two missing arcs, W_1D_3, which was just removed, can give no improvement; as for W_2D_2, it closes the loop $W_2D_2W_1D_1W_2$. We calculate

$$c_{22} - c_{12} + c_{11} - c_{21} = 9 - 4 + 1 - 5 = 1$$

Since this is also positive, it follows that this arc gives no improvement. There are no other arcs. We conclude that this is the solution.

We have seen the simplification of Stage II of the simplex algorithm for the transportation problem. Let us consider next; Stage I: obtaining a b.f.p. A very easy method of obtaining a b.f.p. for these problems is called the *northwest corner* rule. It can be illustrated by referring once again to Example II.5.2.

We wish to obtain a b.f.p. for the problem with availabilities

$$a_1 = 50 \qquad a_2 = 60$$

and requirements

$$b_1 = 30 \qquad b_2 = 40 \qquad b_3 = 40$$

We start at the upper left-hand (i.e., northwest) corner. We have to get 30 units to D_1. Since W_1 has all this (and more) available, we let $x_{11} = 30$. There are still 20 units at W_1. Now D_1 requires 40 units. We give it the 20 remaining units at W_1, and 20 others from W_2, obtaining $x_{12} = 20$, $x_{22} = 20$. There are still 40 units left at W_2; these are needed at D_3 so we let $x_{23} = 40$. We obtain the scheme

	D_1	D_2	D_3
W_1	30	20	0
W_2	0	20	40

which may be seen to be a b.f.p.

This, then, is the general idea behind the northwest corner rule: as many of the D_j as possible are serviced (on a first come, first served basis) from W_1; the remainder are serviced, in turn, from W_2, W_3, and so on. Mathematically, the description is as follows: let x_{11} be the smaller of a_1 and b_1. If $x_{11} < a_{11}$ then let x_{12} be the smaller of $a_1 - x_{11}$ and b_2. If, on the other hand, $x_{11} < b_1$, then let x_{21} be the smaller of a_2 and $b_1 - x_{11}$. Note that each step, in a sense, reduces the problem to that of finding a b.f.p. for a smaller transportation problem. The same process can be continued until all the availabilities and requirements have been exhausted. It is not obvious, but this scheme will actually give a b.f.p. (that is, a vertex of the constraint set, rather than an arbitrary point). The reason is that each of the variables x_{ij} has been found by solving an equation: thus, the point is an intersection of sufficiently many of the bounding planes of the constraint set.

We illustrate this procedure with an example:

II.12.3 Example. Find a b.f.p. for the 6 ×10 transportation program with availabilities

$$\mathbf{a} = (40, 30, 50, 20, 80, 30)$$

and requirements

$$\mathbf{b} = (30, 20, 10, 40, 20, 20, 30, 10, 10, 60)$$

We check first that $\Sigma a_i = \Sigma b_j = 250$.

Using the first warehouse, we service as many of the distributors as possible, starting with D_1. Thus, $a_1 = 40$, and $b_1 = 30$, so we let $x_{11} = 30$. This leaves 10 units at W_1. Since $b_2 = 20$, we set $x_{12} = 10$ and continue to service D_2 from W_2. We get $x_{22} = 10$. There are still 20 units left at W_2; we service D_3 by letting $x_{23} = 10$. The remaining 10 units at W_2 go to D_4 and $x_{24} = 10$. Then we set $x_{34} = 30$ to complete the requirements at D_4. The remaining 20 units at W_3 will go to D_5. We continue in this manner until we obtain the scheme

	D_1	D_2	D_3	D_4	D_5	D_6	D_7	D_8	D_9	D_{10}
W_1	30	10	0	0	0	0	0	0	0	0
W_2	0	10	10	10	0	0	0	0	0	0
W_3	0	0	0	30	20	0	0	0	0	0
W_4	0	0	0	0	0	20	0	0	0	0
W_5	0	0	0	0	0	0	30	10	10	30
W_6	0	0	0	0	0	0	0	0	0	30

It may be checked directly that this scheme satisfies the constraints of the example.

II.12.4. Example. Solve the 3×5 transportation problem with availabilities

$$\mathbf{a} = (50, 45, 65),$$

requirements

$$\mathbf{b} = (20, 40, 40, 35, 25)$$

and unit costs given by the table:

TABLE II.12.2
From Warehouse To Distributor

	D_1	D_2	D_3	D_4	D_5
W_1	2	6	5	3	5
W_2	8	9	7	9	3
W_3	2	3	8	4	6

First of all, we obtain a b.f.p. by the northwest corner method
Thus, W_1 supplies all 20 units for D_1 and 30 units for D_2. W_2 supplies the remaining 10 units for D_2 and 35 units for D_3. W_3 supplies the remainder. We obtain the scheme

	D_1	D_2	D_3	D_4	D_5
W_1	20	30	0	0	0
W_2	0	10	35	0	0
W_3	0	0	5	35	25

We consider, now, the missing Arcs; there are eight of these. The arc W_1D_8 closes the loop $W_1D_8W_2D_1W_1$; accordingly, we calculate

$$c_{13} - c_{23} + c_{22} - c_{12} = 5 - 7 + 9 - 6 = 1$$

Since this is positive, we conclude that the change would not be profitable. We consider other arcs: the arc W_2D_5, finally, is seen to bring a favorable change. In fact, W_2D_5 closes the loop $W_2D_5W_3D_3W_2$; accordingly, we calculate

$$c_{25} - c_{35} + c_{33} - c_{23} = 3 - 6 + 8 - 7 = -2$$

which is negative. Now, $x_{35} = 25$, and $x_{23} = 35$. As $x_{35} < x_{23}$, we remove the arc W_3D_5; we obtain

	D_1	D_2	D_3	D_4	D_5
W_1	20	30	0	0	0
W_2	0	10	10	0	25
W_3	0	0	30	35	0

For this new scheme, we consider the missing arcs. Arc W_3D_1 will close the loop $W_3D_1W_1D_2W_2D_3W_3$. We calculate

$$c_{31} - c_{11} + c_{12} - c_{22} + c_{23} - c_{33} = 2 - 2 + 6 - 9 + 7 - 8 = -4$$

which is negative. We have $x_{11} = 20$, $x_{22} = 10$, and $x_{33} = 30$. As x_{22} is the smallest of these, we shall add the arc W_3D_1 and remove W_2D_2. We now have

	D_1	D_2	D_3	D_4	D_5
W_1	10	40	0	0	0
W_2	0	0	20	0	25
W_3	10	0	20	35	0

For this scheme, the arc W_1D_3 closes the loop $W_1D_3W_3D_1W_1$. We have

$$c_{13} - c_{33} + c_{31} - c_{11} = 5 - 8 + 2 - 2 = -3$$

We have $x_{33} = 20$ and $x_{11} = 10$. Accordingly, we adjoin W_1D_3 and remove W_1D_1, obtaining

	D_1	D_2	D_3	D_4	D_5
W_1	0	40	10	0	0
W_2	0	0	20	0	25
W_3	20	0	10	35	0

The arc W_3D_2 closes the loop $W_3D_2 W_1D_3W_3$. We have

$$c_{32} - c_{12} + c_{13} - c_{33} = 3 - 6 + 5 - 8 = -6$$

We have, also, $x_{12} = 40$ and $x_{33} = 10$. Therefore, we add the arc W_3D_2 and remove W_3D_3. This gives us

	D_1	D_2	D_3	D_4	D_5
W_1	0	30	20	0	0
W_2	0	0	20	0	25
W_3	20	10	0	35	0

For this scheme, W_1D_1 closes the loop $W_1D_1W_3D_2W_1$. We calculate

$$c_{11} - c_{31} + c_{32} - c_{11} = 2 - 2 + 3 - 6 = -3$$

We have $x_{31} = 20$ and $x_{12} = 30$, so we add the arc W_1D_1 and remove W_3D_1. We have, now,

	D_1	D_2	D_3	D_4	D_5
W_1	20	10	20	0	0
W_2	0	0	20	0	25
W_3	0	30	0	35	0

Here, W_1D_4 closes the loop $W_1D_4W_3D_2W_1$. We compute

$$c_{14} - c_{34} + c_{32} - c_{12} = 3 - 4 + 3 - 6 = -4$$

We have $x_{34} = 35$ and $x_{12} = 10$. We therefore add W_1D_4 and delete W_1D_2, which gives us

	D_1	D_2	D_3	D_4	D_5
W_1	20	0	20	10	0
W_2	0	0	20	0	25
W_3	0	40	0	25	0

Now, the arc W_3D_1 closes the loop $W_3D_1W_1D_4W_3$. We calculate

$$c_{31} - c_{11} + c_{14} - c_{34} = 2 - 2 + 3 - 4 = -1$$

We have $x_{11} = 20$, $x_{34} = 25$. Adding W_3D_1 and deleting W_1D_1, we obtain

	D_1	D_2	D_3	D_4	D_5
W_1	0	0	20	30	0
W_2	0	0	20	0	25
W_3	20	40	0	5	0

This scheme is optimal: none of the missing arcs can give an improvement. The total cost, as may be verified directly, is 585, which compares most favorably with the cost of 885, given by the first b.f.p. obtained.

13. ASSIGNMENT PROBLEMS

Example II.12.4 shows the strength of the flow-type algorithm developed by T. C. Koopmans. It should be pointed out that this example has 15 variables; using the simplex algorithm, its tableaux would have eight rows and nine columns each.

Clearly, there is a very real economy of time and effort in using this method. In fact, transportation problems involving some 20 sources and 1000 destinations (and therefore some 20,000 variables) are easily handled in this way with the aid of medium-sized computers.

In case of degeneracy, of course, the Koopmans algorithm fails, because, for degenerate programs, adding a new arc to the graph of a b.f.p. will not necessarily close a loop. If the Program is not greatly degenerate, a perturbation of the program is usually sufficient to solve it; it is enough to put a small flow - an amount ε - on a missing arc that will not close a loop. (Generally, the arc chosen is the one with smallest unit costs among those that do not close a loop.) If, on the other hand, the program is very degenerate - and there are problems that by their very nature give rise to very degenerate programs - then substantially different techniques are necessary. The *assignment problem* is one such.

Let us assume that a company has several positions to be filled and an equal number of job applicants; the only problem is to determine which of the applicants should be given which position. Thanks to aptitude tests, or, perhaps, to some study of the applicants' backgrounds, it is estimated that (for each value of i and j) the company will derive a utility equal to a_{ij} if the ith applicant is assigned to the jth job. The assignment of applicants to jobs is to be made, then, in such a way as to maximize the sum of these expected utilities.

To describe this problem mathematically, we give a definition.

II.13.1. Definition. An $n \times n$ matrix $X = (x_{ij})$ is said to be a *permutation matrix* if all its entries are either 0 or 1, and, moreover, there is exactly one 1 in each row and one 1 in each column.

Thus, the following 5 x 5 matrices are permutation matrices:

(a)

1	0	0	0	0
0	1	0	0	0
0	0	1	0	0
0	0	0	1	0
0	0	0	0	1

(b) $\begin{pmatrix} 1 & 0 & 0 & 0 & 0 \\ 0 & 0 & 0 & 1 & 0 \\ 0 & 1 & 0 & 0 & 0 \\ 0 & 0 & 0 & 0 & 1 \\ 0 & 0 & 1 & 0 & 0 \end{pmatrix}$

(c) $\begin{pmatrix} 0 & 1 & 0 & 0 & 0 \\ 1 & 0 & 0 & 0 & 0 \\ 0 & 0 & 0 & 1 & 0 \\ 0 & 0 & 0 & 0 & 1 \\ 0 & 0 & 1 & 0 & 0 \end{pmatrix}$

It is clear that we can think of each permutation matrix X as an assignment: the ith applicant is assigned to the jth job if $x_{ij} = 1$. Definition II.13.1 guarantees that no person is assigned to two jobs, and that no two persons are assigned to the same job.

We can now express the assignment problem, mathematically. in the form

2.13.1 Maximize

$$\sum_{i=1}^{n} \sum_{j=1}^{n} a_{ij} x_{ij}$$

Subject to

2.13.2 X is a permutation matrix.

Now this is not, strictly speaking, a linear programming problem. In fact the constraint set is finite rather than convex. This type of problem is called a *combinatorial* problem. It would not seem, at first glance, that the assignment problem could be solved by linear programming methods. On the other hand, we should remember that the solution to a linear program (if it exists) can always be found by considering a finite set of points, the extreme points of the constraint set. Hence the assignment problem will reduce to a linear program if we can only choose a constraint set whose extreme points are precisely the permutation matrices.

II.13.2. Definition. An $n \times n$ matrix $X = (x_{ij})$ is said to be *doubly stochastic* if all $x_{ij} \geq 0$ and, moreover,

$$\sum_{j=1}^{n} x_{ij} = 1 \quad \text{for each } i$$

and

$$\sum_{i=1}^{n} x_{ij} = 1 \quad \text{for each } j$$

In essence, a doubly stochastic matrix is a non-negative matrix whose row sums and column sums are all equal to 1. Now, it is clear that any $n \times n$ matrix can be thought of as representing a point in n^2-dimensional space. If so, the set of doubly stochastic matrices forms a convex set, since it is defined solely by linear equations and inequalities. It is easy to see that a permutation matrix is doubly stochastic; the relation between doubly stochastic matrices and permutation matrices is given by the following theorem.

II.13.3. Theorem. The extreme points of the set of doubly stochastic matrices are precisely the permutation matrices.

We will not give a proof of Theorem II.13.3; it may be obtained by an adaptation of the proof of II.12.1. We can now restate the assignment problem in the form

2.13.3 Maximize

$$\sum_{i=1}^{n}\sum_{j=1}^{n} a_{ij} x_{ij}$$

Subject to

2.13.4 $$\sum_{j=1}^{n} x_{ij} = 1 \quad \text{for each } i$$

and

2.13.5 $$\sum_{i=1}^{n} x_{ij} = 1 \quad \text{for each } j$$

2.13.6 $\qquad x_{ij} \geq 0 \qquad$ for each i,j

What we have done is to replace the set of permutation matrices by the set of doubly stochastic matrices. The program (2.13.3) to (2.13.6) is clearly feasible and bounded. It must have a maximum, which is attained at one of the extreme points of the constraint set. This extreme point will be the optimal assignment. It may be seen that program (2.13.3) to (2.13.6) is a transportation-type problem, with availabilities and requirements all equal to 1. Unfortunately, it cannot be handled by the Koopmans algorithm because it is extremely degenerate. in fact, the constraints (2.13.4) and (2.13.5) are $2n$ equations, one of which is redundant. Thus, a b.f.p. should have $2n$-1 non-zero variables. But we know that each b.f.p. has exactly n non-zero variables (one in each row of the matrix). This degeneracy is enough to make the Koopmans algorithm useless.

A common procedure in case of degeneracy is to solve, not the given (primal) linear program, but its dual.

Let us look, then, for the dual of (2.13.3) to (2.13.6). In Section II.11, the dual problem was defined, but only for programs with inequality constraints. This program has equation constraints. As was pointed out earlier, each equation can be replaced by two opposite inequalities: in general, we replace $a = b$ by $-a \geq -b$ and $a \geq b$. Thus, two inequalities arise, in which the coefficients are negatives of each other. The dual program will have one variable for each of these inequalities; these two variables appear, in each of the constraints, with coefficients that are negatives of each other. This being so, we can deal with the difference of the two variables. The two variables are non-negative, but there is no such restriction on their difference. We have, as a rule for duality: *in duality, the variable corresponding to an equation constraint is not restricted to be nonnegative.*

Using this rule, we can see that the dual of program (2.13.3) to (2.13.6) is

2.13.7 $\qquad$ Minimize

$$\sum_{i=1}^{n} u_i + \sum_{j=1}^{n} v_j$$

2.13.8 $\qquad$ Subject to

$$u_i + v_j \geq a_{ij} \quad \text{for each } i, j$$

The dual program attempts to put "shadow prices," u_i, on the job applicants, and v_j on the several jobs. The number u_i represents, in some way, a "natural wage" for the ith applicant while the v_j represents, perhaps, the value of the equipment used for the jth job.

Let us suppose that X^* is an optimal assignment, while $(\mathbf{u}^*, \mathbf{v}^*)$ is a solution to the dual program. We know from duality that this is equivalent to the condition

$$2.13.9 \qquad \sum\sum a_{ij} x^*_{ij} = \sum u^*_i + \sum v^*_j$$

and it is a consequence of complementary slackness that

$$2.13.10 \qquad \text{if } x^*_{ij} = 1, \text{ then } u^*_i + v^*_j = a_{ij}$$

This condition may he interpreted as saying that an employee's salary, plus the value of the job's equipment, should be no greater than his utility to the firm in that job.

Condition (2.13.10) is also sufficient. If there is a permutation matrix X^* such that (2.13.10) holds, then (2.13.9) will also hold, and X^* will be the optimal assignment. Our approach to the assignment problem will be along these lines. We shall look for vectors $(\mathbf{u},\mathbf{v})$ satisfying (2.13.8) and decrease the objective function until a permutation matrix X^* satisfying (2.13.10) is found.

We shall assume that the entries in the matrix $A = (a_{ij})$ are all integers. This is not a great loss in generality, since, whatever the numbers a_{ij} may be, they can be approximated as closely as desired by rational numbers. In turn, these rational numbers may all be multiplied by their least common denominator, thus obtaining an equivalent assignment problem. The following theorem, which is intuitively reasonable, is given without proof.

II.13.4. Theorem. Let $B = (b_{ij})$ be an $n \times n$ matrix, all of whose entries are 0 or 1. A necessary and sufficient condition for the existence of a permutation matrix $X \leq B$ is that every set of k rows have non-zero entries in at least k columns.

Our method of solution of the assignment problem is as follows: A vector $(\mathbf{u},\mathbf{v})$ satisfying (2.13.8) is found. This is most easily done by letting $v_j = 0$, and lettting u_i be equal to the largest entry in the ith row. An auxiliary matrix $B = (b_{ij})$ is then formed, defined by

$$2.13.11 \qquad b_{ij} = \begin{cases} 1 & \text{if } u_i + v_j = a_{ij} \\ 0 & \text{if } u_i + v_j > a_{ij} \end{cases}$$

Note that, by (2.13.10), we can have $x^*_{ij} = 1$ only if $b_{ij} = 1$. If there is a permutation matrix $X \leq B$, we know from the foregoing discussion that this is the optimal assignment. If not, then, by Theorem II.13.4, there is a set of k rows that have 1's in only $k-1$ or fewer columns. Let us say that these k rows form the set I, and the corresponding $k-1$ (or fewer) columns form the set J. We then define a new vector $(\mathbf{u'},\mathbf{v'})$ by

2.13.12
$$u_i = \begin{cases} u_i - 1 & \text{for } i \in I \\ u_i & \text{for } i \notin I \end{cases}$$

2.13.13
$$v_j = \begin{cases} v_j + 1 & \text{for } j \in J \\ v_j & \text{for } j \notin J \end{cases}$$

Consider this vector. We notice, first of all, that $(\mathbf{u'},\mathbf{v'})$ satisfies (2.13.8). In fact, suppose $b_{ij} = 0$. In this case, $a_{ij} < u_i + v_j$. Now, a_{ij}, u_i and v_j are all integers, and so
$$a_{ij} \leq u_i + v_j - 1$$

Now, by (2.13.12), $u_i' = u_i - 1$, and $v_j' \geq v_j$. Thus, $a_{ij} \leq u_i' + v_j'$.

Suppose, on the other hand, that $b_{ij} = 1$. In this case, there are two possibilities. If $i \in I$, then $j \in J$, so $u_i' = u_i - 1$, $v_j' = v_j + 1$, and so $a_{ij} = u_i' + v_j'$. If i is not in I, then $u_i' = u_i$, $v_j' \geq v_j$, and so $a_{ij} \leq u_i' + v_j'$. In any case, we see that $(\mathbf{u'}, \mathbf{v'})$ satisfies (2.13.8).

Consider, next, the objective function (2.13.7). We see that the sum of the u_i decreases by k, while the sum of the v_j increases by at most $k-1$, since there are k rows in I but at most $k - 1$ columns in J. At each step, the objective function (2.13.7) decreases by at least 1. It follows that the minimum of the program (2.13.7) and (2.13.8) will be reached in a finite number of steps; this will give us the optimal assignment. We conclude with an example of this algorithm.

II.13.5. Example. Find the optimal assignment corresponding to the matrix

$$A = \begin{pmatrix} 14 & 18 & 14 & 12 & 19 & 13 & 14 \\ 20 & 17 & 18 & 15 & 12 & 14 & 14 \\ 19 & 18 & 13 & 18 & 17 & 13 & 12 \\ 12 & 20 & 12 & 16 & 13 & 18 & 18 \\ 11 & 16 & 17 & 16 & 11 & 18 & 14 \\ 16 & 14 & 16 & 18 & 20 & 15 & 13 \\ 17 & 12 & 15 & 18 & 13 & 18 & 19 \end{pmatrix}$$

Starting by setting $v_j = 0$ and $\mathbf{u} = (19, 20, 19, 20, 18, 20, 19)$, we obtain the following auxiliary matrix:

$$B = \begin{pmatrix} 0 & 0 & 0 & 0 & 1 & 0 & 0 \\ 1 & 0 & 0 & 0 & 0 & 0 & 0 \\ 1 & 0 & 0 & 0 & 0 & 0 & 0 \\ 0 & 1 & 0 & 0 & 0 & 0 & 0 \\ 0 & 0 & 0 & 0 & 0 & 1 & 0 \\ 0 & 0 & 0 & 0 & 1 & 0 & 0 \\ 0 & 0 & 0 & 0 & 0 & 0 & 1 \end{pmatrix} \quad \mathbf{u} = \begin{pmatrix} 19 \\ 20 \\ 19 \\ 20 \\ 18 \\ 20 \\ 19 \end{pmatrix}$$

$$\mathbf{v} = \begin{pmatrix} 0 & 0 & 0 & 0 & 0 & 0 & 0 \end{pmatrix}$$

We note that the first, second, third and sixth rows of this matrix B have 1's only in the first and fifth columns. We therefore decrease u_1, u_2, u_3, and u_6 by 1, and increase v_1 and v_5:

$$\mathbf{B} = \begin{array}{c} \\ \\ \\ \\ \\ \\ \\ \end{array} \begin{array}{|ccccccc|} \hline 0 & 1 & 0 & 0 & 1 & 0 & 0 \\ 1 & 0 & 0 & 0 & 0 & 0 & 0 \\ 1 & 1 & 0 & 1 & 0 & 0 & 0 \\ 0 & 1 & 0 & 0 & 0 & 0 & 0 \\ 0 & 0 & 0 & 0 & 0 & 1 & 0 \\ 0 & 0 & 0 & 0 & 1 & 0 & 0 \\ 0 & 0 & 0 & 0 & 0 & 0 & 1 \\ \hline \end{array} \quad \mathbf{u} \begin{array}{|c|} \hline 18 \\ 19 \\ 18 \\ 20 \\ 18 \\ 19 \\ 19 \\ \hline \end{array}$$

$$\mathbf{v} = \begin{array}{|ccccccc|} \hline 1 & 0 & 0 & 0 & 1 & 0 & 0 \\ \hline \end{array}$$

This time, we note that first, second, third, fourth and sixth rows of B have 1's only in the first second, fourth, and fifth columns. We decrease u_1, u_2, u_3, u_4, and u_6, and increase v_1, v_2, v_4 and v_5:

$$\mathbf{B} = \begin{array}{|ccccccc|} \hline 0 & 0 & 0 & 0 & 1 & 0 & 0 \\ 1 & 0 & 1 & 0 & 0 & 0 & 0 \\ 1 & 1 & 0 & 1 & 0 & 0 & 0 \\ 0 & 1 & 0 & 0 & 0 & 0 & 0 \\ 0 & 0 & 0 & 0 & 0 & 1 & 0 \\ 0 & 0 & 0 & 0 & 1 & 0 & 0 \\ 0 & 0 & 0 & 0 & 0 & 0 & 1 \\ \hline \end{array} \quad \mathbf{u} \begin{array}{|c|} \hline 17 \\ 18 \\ 17 \\ 19 \\ 18 \\ 18 \\ 19 \\ \hline \end{array}$$

$$\mathbf{v} = \begin{array}{|ccccccc|} \hline 2 & 1 & 0 & 1 & 2 & 0 & 0 \\ \hline \end{array}$$

For this scheme, we note that the first, fourth, and sixth rows of B have 1's only in the second and fifth columns. We will therefore decrease u_1, u_4 and u_6 and increase v_2 and v_5, to obtain the scheme

$$B = \begin{bmatrix} 0 & 1 & 0 & 0 & 1^* & 0 & 0 \\ 1 & 0 & 1^* & 0 & 0 & 0 & 0 \\ 1^* & 0 & 0 & 1 & 0 & 0 & 0 \\ 0 & 1^* & 0 & 0 & 0 & 1 & 1 \\ 0 & 0 & 0 & 0 & 0 & 1^* & 0 \\ 0 & 0 & 0 & 1^* & 1 & 0 & 0 \\ 0 & 0 & 0 & 0 & 0 & 0 & 1^* \end{bmatrix} \quad u = \begin{bmatrix} 16 \\ 18 \\ 17 \\ 18 \\ 18 \\ 17 \\ 19 \end{bmatrix}$$

$$v = \begin{bmatrix} 2 & 2 & 0 & 1 & 3 & 0 & 0 \end{bmatrix}$$

We check, and find that this matrix B does, in fact, contain a permutation matrix; this is shown by the starred elements. The optimal assignment is shown by the starred elements in the matrix A that follows:

$$A = \begin{bmatrix} 14 & 18 & 14 & 12 & 19^* & 13 & 14 \\ 20 & 17 & 18^* & 15 & 12 & 14 & 14 \\ 19^* & 18 & 13 & 18 & 17 & 13 & 12 \\ 12 & 20^* & 12 & 16 & 13 & 18 & 18 \\ 11 & 16 & 17 & 16 & 11 & 18^* & 14 \\ 16 & 14 & 16 & 18^* & 20 & 15 & 13 \\ 17 & 12 & 15 & 18 & 13 & 18 & 19^* \end{bmatrix}$$

It may be checked directly that this is, indeed, the optimal assignment. In fact, we have a value of 131 for the objective functions of both programs; duality assures that these are the solutions.

We point out that this program has 49 variables, constrained by 13 inequalities. Each simplex tableau would have 14 rows and 37 columns (counting the column of constant terms and the objective function row). If we tried, alternatively, to enumerate all the possible assignments, we would find that there are 7! = 5040 such assignments. In either case, it is clear that this algorithm affords us a very substantial economy of time and labor.

Many types of problem can be solved by methods similar to those considered in this section. One such is the flow problem, which is that of maximizing the total flow through a network whose arcs and nodes have finite capacities. It is not, unfortunately, possible to consider all such problems here.

One problem that does not seem soluble by these methods is the so-called "traveling salesman" problem, which consists, essentially, of finding the shortest route through a network, touching all the nodes. Heuristically, it describes the problem faced by a salesman who wishes to visit several cities and wishes to find the shortest (or cheapest) route through all of them. As such, it does not seem greatly different from the assignment problem; it cannot however, be reduced to a linear program without the introduction of a very large number of constraints. No solution has yet been given in general (though some particular problems of this type have been solved).

III. THE THEORY OF PROBABILITY

1. PROBABILITIES

The theory of probability arose from the desire of men to understand the behavior of systems that cannot entirely be controlled. More exactly, it arose from the desire of a French gambler, the Chevalier de Meré, to understand why his dice had suddenly turned against him. The Chevalier's concern was a very practical one: he was losing money.

For some time, de Meré had bet (at even odds) that, in four tosses of a die, he would obtain a six at least once. This was a profitable bet, as was attested to by his systematic winnings Unfortunately, he had been too successful and, as a result, found himself unable to obtain any takers for his bets. This led him to change the bet: he now bet that, in 24 tosses of a pair of dice, he would obtain a double six at least once. The Chevalier reas6ncd that this bet would be as profitable as the previous one (since, indeed, $4/6 = 24/36$); the change in the bet however, would enable him to find new takers. He did find new takers for the bet, but, to his surprise, he began to lose systematically.

We shall consider this example later on, and see what de Meré's mistake consisted in. For the present we will simply note that, not being a mathematician, he proceeded to consult one - Blaise Pascal (1623-1662), one of the greatest of all time. Pascal discussed the problem with Pierre de Fermat (1601-1665) and, from the analysis made by these two men, a new branch of mathematics was born.

The fundamental problem in the theory of probability is, then, as follows. A certain action, called an *experiment*, is carried out. The experiment may have any one of several possible *outcomes;* the idea is to predict, more or less, the frequency with which these outcomes (or certain sets of them, called *events)* will appear - assuming that the experiment can be repeated many times. The experiment can be of many types: we give examples.

(a) A die is tossed: the *outcome* is the face that lands on the *bottom*.
(b) A dart is thrown at a bull's-eye; the *outcome* is the section of the bull's-eye in which the dart lands.
(c) A coin is tossed three times; the *outcome* is the run of heads and tails obtained.
(d) A telephone book is opened; the *outcome* is the page to which it is opened.

We give these examples to illustrate several things. Example (a) shows that the outcomes may be labeled in many ways (since it is usual to think of the *top* face of the die as the outcome of this experiment). Example (b) shows that many outcomes may be classed together as a single outcome (we could consider each point on die board as a different outcome). Example (c) shows that an experiment may consist of several other, simpler experiments. Example (d) shows that the experimenter may be able to influence, to some degree, the outcome, since he can choose, to some extent the section of die book to which he will open (though possibly not the exact page). It is often necessary to explain the manner in which the experiment is carried out.

Now, how are probabilities to be assigned? This is an extremely complicated question - not from a mathematical, but from a practical point of view. Mathematically, we can give a system of axioms that are to be satisfied by any assignment of probabilities. The difficulty lies in our practical desire that the probability assigned to an outcome correspond to the relative frequency with which it occurs. This can be ascertained only through experimentation. But such experiments, precisely because of their nature, are always subject to random fluctuations, which tend to, but need not, cancel out over the long run. Moreover, the "long run" might be so long as to be quite impractical.

An example may be of use in explaining this difficulty. A coin is said to be *fair* if, when it is tossed, each of the two outcomes (heads and tails) has a probability 1/2 of occurring. This means (for reasons to be seen later) that, in tossing a fair coin 20 times, the probability of obtaining heads every time is $(1/2)^{20}$, or approximately 0.000001. Thus, if a coin is tossed 20 times, and turns up heads every time, we would be justified in claiming that the coin is skew (i.e., not fair).

Suppose, however, that each adult person in the United States (say, some 150,000,000 people) were to toss a fair coin 20 consecutive times. In this case, it is almost certain that approximately 150 of them would have the singular experience of seeing a string of 20 consecutive heads; another 100 would obtain a string of 20 consecutive tails. We might imagine each of these 200 claiming that the coins they had used were entirely skew; in fact however, we would claim that it was a cause for surprise if no such strings of 20 heads occurred. The idea is that, no matter how unlikely an outcome, it is likely to happen if sufficiently many repeated attempts are made, and so we have to be careful in assigning probabilities on the basis of observed results.

A second conceptual difficulty is the fact that in many cases, the experiment is such that it cannot or will not be repeated. The Sox meet the Cats tomorrow to determine the Baseball Championship; what is the probability that the Sox will win? A building contractor makes a bid on a projected building; what is the probability that he will be the low bidder? Mr. Jones will go hunting tomorrow for the first time in his life; what is the probability that he will shoot a fox? In each of these cases, it is clear that the "experiment" with all its implications, cannot be repeated at all (the baseball players, for example, could play a seven game championship set, but then the individual games would no longer be as important as the single championship game). Thus a probability can be assigned to these outcomes only by defining the concept of probability in a different manner.

In such cases, it is common to talk about *subjective* probability. The subjective probability of an event is, more or less, the belief that an individual has as to its likelihood; the event has, say, a subjective probability of 1/2 if an individual feels that a bet at even odds is a "fair" bet. This is not, to say the least, a very scientific idea; it is further complicated by the fact that many people may have different beliefs about the event's likelihood, and so the subjective probability is not well defined. Practical men have, of course, found a way to solve this problem. The bookmaker, seeing that he cannot reconcile the subjective probabilities held by different people, looks for an equilibrium value (for the odds he offers) that will guarantee that he wins whether, in fact, the Sox or the Cats win the championship.

The problem of assigning probabilities to the outcomes of an experiment remains. In the absence of anything better, the rule of Pierre Simon, Marquis de Laplace (1749-1827), has been used: *when there is no indication to the contrary, assume that each of the possible outcomes has the same probability*. When tossing a coin, one assumes (unless there is some indication to the contrary) that each of the two sides has the same probability. If an urn has 50 balls, all identical except for color, then we assume that each of the balls has the same probability of being drawn nut by a blindfolded man. Laplace's rule does not hold when it seems reasonable that one of the outcomes is more likely than another. For instance, if a die has one face much larger be others, it seems reasonable to assume that the die will land on that face more often than on the others.

In the rest of this chapter, we shall assume that probabilities can be assigned to the outcomes; we shall then attempt to draw logical conclusions about the outcomes of other, more complicated experiments. This is the domain of probability theory. The problem of assigning these original probabilities belongs to the field of statistical analysis and is treated in books on statistics.

2. DISCRETE PROBABILITY SPACES

We shall begin our study of probability by considering only experiments with a finite number of possible outcomes.

III.2.1. Definition. A finite probability space is a finite set of outcomes,

$$X = \{O1, O2, ..., On\},$$

together with a function, P, which assigns a non-negative real number to each outcome, in such a way that

3.2.1 $$\sum_{i=1}^{n} P(Oi) = 1.$$

The number $P(Oi)$ is the probability of the outcome Oi; it is often replaced by p_i. The reason for condition (3.2.1) should be clear: since each trial of the experiment will have one (and only one) of the outcomes Oi, it follows that the sum of their relative frequencies must be equal to 1.

III.2.2 Definition. An event is a subset of the set X.

An event is defined as a set of outcomes. The probability function P, which is defined for outcomes, can be extended to events by saying that an event happens whenever one of the outcomes belonging to this event occurs. Then, the probability of the event will be the sum of the probabilities of its several outcomes.

3.2.2 $$P(A) = \sum_{Oi \in A} P(Oi)$$

In particular, X itself is an event called the certain event, since it contains all the possible outcomes. The empty set, $\emptyset$, is also an event, called the impossible event. It is easy to see that

3.2.3 $$P(X) = 1$$

3.2.4 $$P(\emptyset) = 0$$

A certain event must have probability 1; an impossible event, probability 0. The converse of this is not quite true; an event may be possible and still have probability 0 or it may have probability 1 and still not be certain. For finite probability spaces, however, this will not happen.

We give now examples of finite probability spaces.

III.2.3. Example. Consider the experiment of tossing a die and looking at the top face. The experiment has six possible outcomes:

$$X = \{1, 2, 3, 4, 5, 6\}$$

If the die is fair, then each of these outcomes will have the same probability; since the sum of these must be 1, it follows that

$$p_i = 1/6 \qquad \text{for } i = 1, 2, 3, 4, 5, 6$$

Consider the event "even number":

$$E = \{2, 4, 6\}$$

We can calculate that

$$P(E) = p_2 + p_4 + p_6 = 1/2$$

Similarly, if $S = \{1, 2\}$, we have

$$P(S) = p_1 + p_2 = 1/3$$

III.2.4. Example. Let the experiment consist of three tosses of a coin; the outcome will be the run of heads and tails obtained. There are 8 possible outcomes:

HHH	*THH*
HHT	*THT*
HTH	*TTH*
HTT	*TTT*

If the coin is fair, each of these outcomes will have the same probability, namely, 1/8. If we consider the event "two heads" we have

$$E = \{HHT, HTH, THH\}$$

and it follows that

$$P(E) = 3/8$$

III.2.5. Example. Let the experiment consist of drawing a card from a deck. There are 52 possible outcomes; the event "hearts" will contain 13 of these outcomes. If we assume that each card has the same probability, 1/52, of being drawn, then

$$P(\text{hearts}) = 13/52 = 1/4$$

For the same experiment, the event "ace" will contain four of the possible outcomes. Thus

$$P(\text{ace}) = 4/52 = 1/13$$

It is not too difficult to see that, if A and B are disjoint events, i.e., if $A \cap B = \emptyset$, then

3.2.5 $$P(A \cup B) = P(A) + P(B)$$

In fact, $P(A \cup B)$ is the sum of the probabilities of the outcomes that belong to $A \cup B$. This sum can be split into two sums, corresponding to $P(A)$ and $P(B)$. Note that this would not be possible if $A \cap B$ were non-empty, since then any outcome that belonged to $A \cap B$ would be counted twice.

If we have several disjoint events, i.e., events $A_1, A_2, \ldots A_m$, such that $A_j \cap A_k = \emptyset$ whenever $j \neq k$, then (3.2.5) generalizes directly, to give

3.2.6 $$P(A_1 \cup A_2 \cup \ldots \cup A_m) = P(A_1) + P(A_2) + \ldots + P(A_m)$$

Again, we point out that (3.2.6) will not hold unless the events A_k are disjoint. In case two events are not disjoint, we have the more general relation

3.2.7 $$P(A \cup B) = P(A) + P(B) - P(A \cap B)$$

To prove (3.2.7), note that the three events

$$A \cap B \qquad A-B \qquad B-A$$

are disjoint. Now

FIGURE III.2.1. Dissection of the set A∪B into the disjoint subsets A-B, B-A, and A∩B.

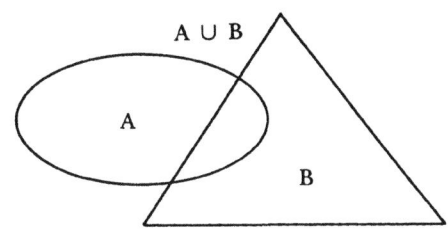

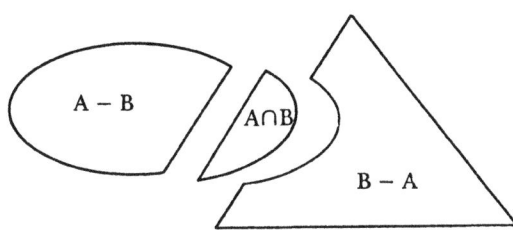

$$A \cup B = (A \cap B) \cup (A\text{-}B) \cup (B\text{-}A)$$

(See Figure III.2.1.) Thus by (3.2.6),

3.2.8 $\qquad P(A \cup B) = P(A \cap B) + P(A\text{-}B) + P(B\text{-}A)$

We also have

$$A = (A \cap B) \cup (A\text{-}B)$$

and

$$B = (A \cap B) \cup (B\text{-}A)$$

so that

3.2.9 $\qquad P(A) = P(A \cap B) + P(A\text{-}B)$

3.2.10 $\qquad P(B) = P(A \cap B) + P(B\text{-}A)$

Adding (3.2.9) and (3.2.10), and substituting (3.2.8), we obtain

$$P(A) + P(B) = P(A \cup B) + P(A \cap B)$$

which is equivalent to (3.2.7).

III.2.6. Example. Given the same experiment as in Example III.2.5, find the probability of obtaining *either* a heart *or* an ace.

For this problem, we can write

$$H = \text{"hearts"}$$
$$A = \text{"ace"}$$

and we are looking for $P(H \cup A)$. We already know $P(H)$ and $P(A)$; to find $P(H \cap A)$, note that $H \cap A$ consists of one single outcome (the ace of hearts) and so

$$P(H \cap A) = 1/52$$

Applying (3.2.7), now, we will have

$$P(H \cup A) = P(H) + P(A) - P(H \cap A) = 1/4 + 1/13 - 1/52 = 16/52$$

and so

$$P(H \cup A) = 16/52 = 4/13$$

This could, of course, have been obtained directly by noticing that there are exactly 16 cards that are either hearts or aces: 13 hearts and 3 other aces.

III.2.7. Example. Let an experiment consist of a toss of two dice. There are, all told, 36 possible outcomes (assuming that the two dice are distinguishable):

$$(1,1), (1,2), (1,3), (1,4), (1,5), (1,6)$$
$$(2,1), (2,2), (2,3), (2,4), (2,5), (2,6)$$
$$(3,1), (3,2), (3,3), (3,4), (3,5), (3,6)$$
$$(4,1), (4,2), (4,3), (4,4), (4,5), (4,6)$$
$$(5,1), (5,2), (5,3), (5,4), (5,5), (5,6)$$
$$(6,1), (6,2), (6,3), (6,4), (6,5), (6,6)$$

Consider the event A: "the sum of the numbers on the two dice is equal to 10." We have

$$A = \{(6,4), (5,5), (4,6)\}$$

and so, assuming that each outcome has probability 1/36, we find

$$P(A) = 3/36 = 1/12$$

In a similar way, the event B, "doubles are thrown," will contain six outcomes:

$$B = \{(1,1), (2,2), (3,3), (4,4), (5,5), (6,6)\}$$

and so

$$P(B) = 6/36 = 1/6$$

We can see also that $A \cap B = (5,5)$; hence

$$P(A \cap B) = 1/36$$

and so

$$P(A \cup B) = 3/36 + 6/36 - 1/36 = 8/36 = 2/9$$

III.2.8. Example. There are 39 families in a small town: of these, 23 are Republican and 16 are Democrat. There are 7 families headed by a woman, and of these, 5 are Democrat. A poll sampler chooses one family at random. Find the probability that: (a) the family is Democrat, (b) it is headed by a woman, (c) it is *either* Democrat *or* headed by a woman.

Let D be the event "the family is Democrat," and let W be the event "It is headed by a woman." Assuming that each family has a probability 1/39 of being chosen, we have:

$$P(D) = 16/39$$

$$P(W) = 7/39$$

while for the joint event, "Democratic *and* headed by a woman",

$$P(D \cap W) = 5/39$$

We thus have

$$P(D \cup W) = 16/39 + 7/39 - 5/39 = 18/39$$

so that the probability that the family is *either* Democrat *or* headed by a woman is 18/39, or 6/13.

PROBLEMS ON DISCRETE PROBABILITY SPACES

1. A fair die is tossed. Let E be the event $\{2,4,6\}$ and let S be the event $\{1,2\}$. Find the probabilities of the events E, S, $E \cap S$, $E-S$, $E \cup S$, and S-E.

2. A fair coin is tossed three times. List the eight possible outcomes. Find the probabilities of the events:

(a) The first toss is H.
(b) Either the first toss is H, or the third toss is T.
(c) Both the second and third tosses are T.
(d) No tosses are H.
(e) At least one of the tosses is T.
(f) The first toss is H, and the third toss is T.
(g) There are at least two T's

3. A card is drawn at random from a deck (of 52 cards). What is the probability that;

(a) The card is an ace?
(b) It is the ace of hearts?
(c) It is *either* an ace *or* a heart?

4. A pair of fair dice is tossed. What is the probability that:

(a) The sum of the spots showing on the top of the dice is 8?
(b) It is 8 or more?
(c) It is 4 or less?

3. CONDITIONAL PROBABILITY

As was mentioned before, we are interested in finding relationships among the probabilities of various events in order to analyze more complicated experiments. Let us consider the idea of *conditional* probability: given that an event A happens, what is the probability of a second event, B?

From a practical point of view, probability theory is applicable when the analyst lacks information about all the relevant variables. A poker player must rely on the theory of probability because he does not know what the card on top of the deck is; if he could peek, he would know whether to bet or fold. Suppose, however, that he notices the top card has a frayed corner. If he knows this deck of cards, the knowledge gives him additional information, which will make his decision easier for him. He no longer has to consider all the cards in the deck, but only those that have frayed corners.

To come back to our original question, the probability of event B, given that event A happens, is the relative frequency of B, considering only those cases in which A happens. This leads to the following definition:

III.3.1. Definition. The *conditional probability* of the event B, given the event A, is the probability $P(B|A)$, given by

3.3.1
$$P(B|A) = \frac{P(A \cap B)}{P(A)}$$

III.3.2. Example. In the experiment of Example III.2.7, find the conditional probability of obtaining doubles, given that the sum of the numbers on the two dice is 10.

In this case, we are looking for $P(B|A)$. We have, from III.2.7,

$$P(A \cap B) = 1/36$$

$$P(A) = 1/12$$

Therefore

$$P(B|A) = (1/36)/(1/12) = 1/3$$

Note that $P(B|A)$ is twice $P(B)$. Whether A occurs will alter the probability of B.

III.3.3. Example. Using the data of Example III.2.8, find the probability that the family sampled is Democrat, given that it is headed by a woman. Find also the probability that it is headed by a woman, given that it is Democrat. (Note that these are not the same thing.)

We are looking here for $P(D|W)$ and $P(W|D)$. Now,

$$P(D|W) = P(D \cap W) / P(W) = (5/39)/(7/39) = 5/7$$

while

$$P(W|D) = P(W \cap D) / P(D) = (5/39)/(16/39) = 5/16$$

Two events are thought of as independent if neither has any influence on the other. Probabilistically, we can best express this by saying that the conditional probability is the same as the absolute (i.e., not conditional) probability.

III.3.4. Definition. Two events, A and B, are independent if

3.3.2 $$P(A \cap B) = P(A)P(B)$$

For independent events, the *joint probability* (i.e., the probability $P(A \cap B)$ that they will both happen) is equal to the product of their individual absolute probabilities.

III.3.5. Example. Consider the experiment of Example III.2.4. Let A be the event "the first toss is a head, and let B be the event "the third toss is tails." We find then that

$$A = \{HHH, HHT, HTH, HTT\}$$

$$B = \{HHT, HTT, THT, TTT\}$$

and

$$A \cap B = \{HHT, HTT\}$$

Since each outcome has probability 1/8, we find that

$$P(A) = P(S) = 1/2$$

$$P(A \cap B) = 1/4 = (1/2)(1/2)$$

The joint probability is the product of the absolute probabilities, and we conclude that the two events are independent. This corresponds to the intuitive idea that the result of the first toss should have no influence on the third toss, or vice-versa. On the other hand, note that for Example III.2.8, the events D and W are *not* independent

For more than two events, independence is defined in a slightly different manner:

III.3.6. Definition. Three events A, B, and C are independent if any two of them are independent and, moreover,

$$P(A \cap B \cap C) = P(A)P(B)P(C)$$

In general, n events $A_1, A_2, \ldots, A_n$ are independent if any $n-1$ are independent and also

3.3.4 $\qquad P(A_1 \cap A_2 \cap \ldots \cap A_n) = P(A_1)P(A_2)\ldots P(A_n)$

Definition III.3.6 is an inductive definition: it defines independence for n events in terms of independence for $n-1$ events, plus an additional condition. An alternative definition would be to say that a collection of events is independent if, for any subcollection, the joint probability is the product of the absolute probabilities.

Note that it is possible for three events not to be independent although any two of them are independent. (Conversely, three events may satisfy (3.3.3) and still not be independent because two of them are not.)

III.3.7. Example. In the experiment of Example III.2.7, let A be the event "the first die shows a 6"; let B be the event "the second die shows a 3" and let C be the event "the sum of the numbers on the two dice is 7." We have, then,

$$A = \{(6,1), (6,2), (6,3), (6,4), (6,5), (6,6)\}$$

$$B = \{(1,3), (2,3), (3,3), (4,3), (5,3), (6,3)\}$$

$$C = \{(1,6), (2,5), (3,4), (4,3), (5,2), (6,1)\}$$

and so $P(A) = P(B) = P(C) = 1/6$. To check for independence, we see that

$$A \cap B = \{(6,3)\}$$

$$A \cap C = \{(6,1)\}$$

$$B \cap C = \{(4,3)\}$$

and so

$$P(A \cap B) = 1/36 = P(A)P(B)$$

$$P(A \cap C) = 1/36 = P(A)P(C)$$

$$P(B \cap C) = 1/36 = P(B)P(C)$$

We see that any two of these events are independent. On the other hand,

$$A \cap B \cap C = \emptyset$$

so that $P(A \cap B \cap C) = 0$, and it follows that the three events are not independent.

4. COMPOUND EXPERIMENTS

In the last few examples, we have assumed that we are given the joint probabilities of the events directly, i.e., that we have been given the probabilities of each of the outcomes. In general, however, this is not so: the problem usually involves computing the probabilities of the outcomes.

Let us now, for instance, consider the problem of the Chevalier de Méré. The Chevalier's original bet was: that he would obtain at least one six in four tosses of a die. We can find the probability of this event by considering the complementary event, "no sixes are obtained." Assuming the die to be fair, the probability of not obtaining a six on a given toss is 5/6; since the four tosses are independent, we must multiply four such factors, i.e.,

$$P(\text{no sixes are obtained}) = (5/6)^4 = 625/1296$$

And so the Chevalier had a probability of $1 - 625/1296 = 671/1296$, or slightly more than 1/2, of winning. He had an advantage: in the long run, he would win more often than not[1]

Consider now the Chevalier's second bet: the probability of not obtaining a double six on any given toss of the dice, assuming the dice to be fair, is 35/36. For 24 independent tosses, we have

$$P(\text{no double sixes are obtained}) = (35/36)^{24}$$

This probability, computed by using logarithms (or a computer), is approximately 0.512. We conclude that de Méré had only a probability 0.488 of winning. In the long run, he should lose somewhat more often than he won - a prediction verified by his experience. The two bets were by no means equivalent. Some mathematical analysis would have saved the Chevalier a great sum of money - but, as Pascal pointed out, de Méré was no mathematician.

III.4.1. Example. A test consists of five multiple-choice questions, each one giving a choice of four possible answers. A student who has not read the subject matter guesses at each answer. What is the probability that he will answer all five questions correctly?

In a sense, this problem is not entirely well defined. The difficulty is that, even though a student has not read the subject matter, the correct answer may yet be somewhat more logical than any of the others, so that an "educated guess" gives a greater probability of success than a wild guess. It may also be that the five questions are logically related, so that the student's success on one of them is not independent of success on the others. Notwithstanding these possibilities, we shall assume that the student has a 1/4 probability of success on each question (since there are four answers, of which only one is correct), and that the several questions are independent. Letting S_i be the event "success on the ith question," we have

$$P(S_i) = 1/4 \qquad \text{for } i = 1, 2, 3, 4, 5$$

We are looking for the probability of success on all the questions, i.e. for $P(S_1 \cap S_2 \cap S_3 \cap S_4 \cap S_5)$. Since the questions are assumed independent, we have, by (3.3.4),

$$P(S_1 \cap S_2 \cap S_3 \cap S_4 \cap S_5) = (1/4)^5 = 1/1024$$

When events are not independent, we are quite often given their conditional probabilities. In such cases, we find joint probabilities by rewriting (3.3.1) in the form

3.4.1 $$P(A \cap B) = P(A) P(B|A)$$

III.4.2. Example. Three identical urns are labeled A, B, and C. Urn A contains 5 red and 7 black balls. Urn B contains 1 red and 5 black balls. Urn C contains 12 red and 3 black balls. An urn is chosen at random, and then a ball is taken at random from the urn. What is the probability that the ball will be red?

Two assumptions are made here: the first is that each urn has the same probability, 1/3, of being chosen. The second is that, from a given urn, each of the balls has the same probability of being chosen. Thus, the conditional probability of a red ball, given that urn A is chosen, will be 5/12 (since 5 of the 12 balls in the urn are red); it is similar for the other urns. Letting R be the event "a red ball is chosen," we find that

$$P(A) = P(B) = P(C) = 1/3$$

$$P(R|A) = 5/12$$

$$P(R|B) = 1/6$$

$$P(R|C) = 4/5$$

The event R can be written as the union of disjoint events

$$R = (R \cap A) \cup (R \cap B) \cup (R \cap C)$$

and so, by (3.2.6),

$$P(R) = P(R \cap A) + P(R \cap B) + P(R \cap C)$$

Now, by (3.4.1),

$$P(R \cap A) = P(A)P(R|A) = (1/3)(5/12) = 5/36$$

$$P(R \cap B) = P(R)P(R|B) = (1/3)(1/6) = 1/18$$
$$P(R \cap C) = P(C)P(R|C) = (1/3)(4/5) = 4/15$$

so that

$$P(R) = 5/36 + 1/18 + 4/15 = 83/180$$

III.4.2. Example. A lot of 20 flashbulbs contains 17 good and 3 defective items. A sample of size 3 is taken. What is the probability that all 3 will be good?

We assume that each item in the lot has the same probability of being chosen. Let *I* be the event "the first item chosen is good", *II* be the event "the second item is good," and *III* be the event "the third item is good." We will apply (3.4.1) twice:

$$P(I \cap II \cap III) = P(I \cap II) P(III|I \cap II) = P(I) P(II|I) P(III|I \cap II)$$

Since 3 of the 20 bulbs are defective, we have, clearly,

$$P(I) = 17/20$$

Suppose that the first item is good. In this case, there will be only 19 bulbs left, including 16 good ones, and hence,

$$P(II|I) = 16/19$$

Similarly, if the first two bulbs chosen are good there are 15 good bulbs among the 18 left, and so

$$P(III|I \cap II) = 15/18$$

Thus,

$$P(I \cap II \cap III) = (17/20)(16/19)(15/18) = 34/57$$

III.4.3. Example. Find the probability that, from a group of 24 people, no 2 have the same birthday.

For this problem, we shall assume that each day is as likely to be a birthday as any other. Let A_i be the event "the *i*th person does not have the same birthday as any of the first i-1." Then we are looking for the probability of the event $A_1 \cap A_2 \cap \ldots \cap A_n$, where n is the number of people in the group; in particular we are interested in the case $n = 24$.

Suppose, now, that all the events $A_1, A_2, \ldots A_{i-1}$ occur. This means that no two of the first i-1 persons have a common birthday; then they occupy i-1 different days. For event A_i to occur, the ith person must then be born in one of the remaining 365 - (i-1) days. Thus,

$$P(A_i | A_1 \cap \ldots \cap A_{i-1}) = (366-i)/365$$

Applying rule (3.3.4), now,

$$P(A_1 \cap \ldots \cap A_n) = (365/365)(364/365)\ldots(366-n/365)$$

For $n = 24$, we have

$$P(A_1 \cap \ldots \cap A_{24}) = (365/365)(364/365)\ldots(342/365) = 0.46$$

We see, thus, that the probability that no 2 have the same birthday is *less than* 1/2! This means that, from 24 people, 2 will have a common birthday more often than not - a strangely unintuitive result, since, indeed, 24 is much smaller than 365.

PROBLEMS ON CONDITIONAL PROBABILITIES

1. A pair of fair dice is tossed. What is the probability that one of the dice show three spots, given that the sum of the spots is 5?

2. A fair coin is tossed three times. What is the probability that the first toss will be H, given that there are at least two H's thrown?

3. A test consists of five multiple-choice questions; each question has three possible answers. A student guesses at all the answers; what is the probability that he will obtain a grade of 20 per cent or better? of 100 per cent?

4. Three otherwise identical urns are labeled *A*, *B*, and *C*. Urn *A* contains 5 red and 6 black balls. Urn *B* contains 3 red and 1 black ball. Urn *C* contains 10 red and 20 black balls. An urn is chosen at random (probability 1/3 each) and then a ball is taken from the urn. What is the probability that it will be red?

5. A chewing gum company has 30 different baseball cards, and gives 1 with each pack of gum. A boy buys 10 packs of gum. What is the probability that he will obtain 10 different cards?

6. A machine makes bolts according to specifications. When it is working well, the bolts meet the specifications with probability 0.98. However, there is a probability 0.35 that the machine will be working poorly, in which case the bolts will meet specifications with probability 0.40. What is the probability that a given bolt will meet the specifications?

7. The game of *craps* is played in the following manner: a player tosses a pair of dice. If the number obtained is 2, 3, or 12, he loses immediately; if it is 7 or 11, he wins immediately. If any other number is obtained on the first toss, then that number becomes the player's "point," and he must keep on tossing the dice until either he "makes his point" (i.e., obtains the first number again), in which case he wins; or he obtains 7, in which case he loses. Find the probability of winning.

8. A machine makes pieces to fit specifications; the probability that a given piece will be suitable is 0.95. A sample of 10 pieces is taken from the machine's production. What is the probability that all 10 pieces will be good?

9. A traveling salesman assigns 0.3 as the probability of making a sale at each call. How many calls must he make in order that the probability of making at least one sale be 0.9 or better?

10. Three urns are labeled *A, B,* and *C.* Urn *A* has four red and two black balls Urn *B* has five red and seven black balls; urn *C* has two red and six black balls An urn is chosen at random and a ball is taken from that urn. What is the probability that the ball will be black?

5. BAYES' FORMULA

A very common application of conditional probabilities deals with what is known as *a priori* and *a posteriori* probabilities.

As an illustration of this type of problem, let us consider the experiment of Example III.4.2. We know that each of the three urns has the same probability, 1/3, of being chosen. Suppose that the three urns are indistinguishable; we could, nevertheless, find out which is which by the simple expedient of looking inside the urn (the number of balls of each color will determine which urn, i.e. if we find that it has 12 red and 3 black balls then we know that it must be urn C). Unfortunately, it may not be possible to look at all the balls in the urn; the question is whether seeing just one of them will give us any information. In fact, this generally does happen. An extreme example would be the case in which urns A and B had no red balls; a red ball would then, necessarily, be drawn from urn C. In our example, the situation is not so extreme, but still the contents of the three urns are sufficiently different so that some new information is gained. Since urn C has such a high proportion of red balls (12 out of 15), a red ball would most likely be drawn from urn C. In fact, we have

$$P(R \cap C) = P(C)P(R|C) = 4/15$$

and

$$P(R) = 83/180$$

Therefore

$$P(C|R) = (4/15)/(83/180) = 48/83$$

so that the *conditional* probability is considerably higher than the absolute probability $P(C)$.

In general, let us assume that the outcomes of an experiment can be classified in two ways (as in our example, by urn and by color). This gives us two partitions of the outcome space:

$$X = A_1 \cup A_2 \cup ... \cup A_n$$

$$X = B_1 \cup B_2 \cup ... \cup B_m$$

It is assumed that, for the first classification, the absolute probabilities $P(A_i)$ are known. For the second partition, it is the *conditional* probabilities $P(B_j|A_i)$ that are known. Now the nature of the experiment is such that it is easy to observe the value of the B_j, whereas the value of the A_i is either impossible or extremely difficult to ascertain. The idea is then to use the observed value B_j to obtain a new (conditional) probability $P(A_i|B_j)$.

This new distribution of the A_i is known as the *a posteriori* distribution while the previous (absolute) distribution is known as the *a priori* distribution. In our example, each of the three urns had an *a priori* probability 1/3 of being chosen. After the event, i.e., assuming that a red ball has been drawn, we conclude that urn C has a considerably larger *a posteriori* probability, 48/83.

FIGURE III.5.1. Scheme of two-classification experiment.

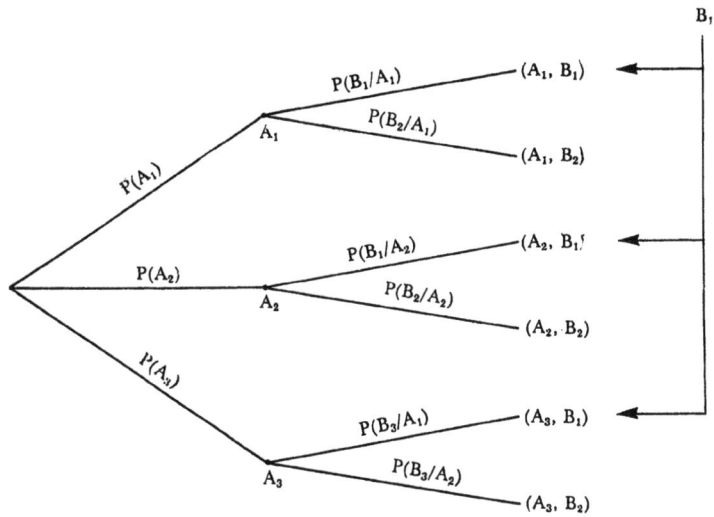

In general, we wish to compute $P(A_i|B_j)$. Now by Definition III.3.1,

$$P(A_i|B_j) = P(A_i \cap B_j)/P(B_j)$$

However, we are given neither $P(A_i \cap B_j)$, nor $P(B_j)$; these must be computed. The numerator presents no trouble; it is given by

$$P(A_i \cap B_j) = P(A_i)P(B_j|A_i)$$

The denominator, on the other hand, requires some work. To simplify things somewhat, let us assume that $m = 2$; i.e., the A classification can give only two values, A_1 and A_2. We can write

$$B_j = (A_1 \cap B_j) \cup (A_2 \cap B_j)$$

Since A_1 and A_2 are disjoint, it follows that $A_1 \cap B_j$ and $A_2 \cap B$ are also disjoint Thus

$$P(B_j) = P(A_1 \cap B_j) + P(A_2 \cap B_j)$$

and, by (3.3.3),

$$P(B_j) = P(A_1)P(B_j|A_1) + P(A_2)P(B_j|A_2)$$

For the special case in which $m = 2$, then, we have the formula

$$P(A_i|B_j) = \frac{P(A_i)P(B_j|A_i)}{P(A_1)P(B_j|A_1) + P(A_2)P(B_j|A_2)}$$

In the more general case, the Formula for $P(A_i|B_j)$ is derived analogously; it is

3.5.1 $$P(A_i|B_j) = \frac{P(A_i) P(B_j|A_i)}{\sum_k P(A_k) P(B_j|A_k)}$$

Equation (3.5.1) is known as *Bayes' formula*.

III.5.1. Example. Of all applicants for a job, it is felt that 75 per cent are able to do the job, and 25 per cent are not. To aid in the selection process, an aptitude test is designed such that a capable applicant has probability 0.8 of passing the test, while an incapable applicant has probability 0.4 of passing the test. An applicant passes. What is the probability that he will be able to do the job?

This is a straightforward application of Bayes' formula. Let A_1 be the event "the applicant is able to do the job," and let A_2 be the complementary event. Let B_1 be the event "the applicant passes the test" and let B_2 the complement. Then

$$P(A_1) = 0.75 \qquad P(A_2) = 0.25$$
$$P(B_1|A_1) = 0.8 \qquad P(B_1|A_2) = 0.4$$

By Bayes' formula,

$$P(A_1|B_1) = \frac{(0.75)(0.8)}{(0.75)(0.8) + (0.25)(0.4)} = \frac{0.6}{0.6 + 0.1} = 0.86$$

The test is a reasonably effective screening device in this case; only 1/7 of the persons passing the test will be incapable.

We may also check for the probability $P(A_1|B_2)$, i.e., the probability that an applicant who fails the test would still be able to do the job. We have

$$P(B_2|A_1) = 0.2 \qquad\qquad P(B_2|A_2) = 0.6$$

And so

$$P(A_1|B_2) = \frac{(0.75)(0.2)}{(0.75)(0.2) + (0.25)(0.6)} = \frac{0.15}{0.15+0.15} = 0.5$$

Thus, half the rejected applicants are able to do the work. Note that the test is not infallible: incapable applicants may pass, while capable applicants may fail.

III.5.2. Example. A machine produces bolts to meet certain specifications. When the machine is in good working order, 90 per cent of the bolts meet the specifications. Occasionally, however, the machine goes out of order, and only 50 per cent of the bolts produced then meet the specifications. It is calculated that the machine is in good order 80 per cent of the time. A sample of two bolts is taken from the machine's current production, and both are found to meet the specifications. What is the probability that the machine is out of order?

Once again, we must use Bayes' formula. Let A_1 be the event "the machine is in good order", and A_2 the complementary event. Let B_1 be the event "both bolts are good." We know that

$$P(A_1) = 0.8 \qquad\qquad P(A_2) = 0.2$$

Next we must compute $P(B_1|A_1)$. If the machine is working well, the probability that a bolt meet specifications is 0.9. The probability that both bolts meet the specifications is, then, $(0.9)^2 = 0.81$.

$$P(B_1|A_1) = 0.81$$

And similarly, $(0.5)^2 = 0.25$, so

Thus,
$$P(B_1|A_2) = 0.25$$

$$P(A_1|B_1) = \frac{(0.8)(0.81)}{(0.8)(0.81)+(0.2)(0.25)} + \frac{0.648}{0.648+0.05} = 0.93$$

III.5.3. Example. Under the same conditions as in Example III.5.2, a sample of two bolts is taken and both are found defective. What is the probability that the machine is out of order?

Let B_2 represent the event "both bolts are defective." As in the previous example, we compute

$$P(B_2|A_1) = (0.1)^2 = 0.01$$

$$P(B_2|A_2) = (0.5)^2 = 0.25$$

Applying Bayes' formula, we have

$$P(A_2|B_2) = \frac{(0.2)(0.25)}{(0.8)(0.01)+(0.2)(0.25)} + \frac{0.05}{0.008+0.05} = 0.86$$

III.5.4. Example. A person is either suffering from tuberculosis (with probability 0.01) or not. To test for this disease, chest x-rays are taken. If the person is ill, the test will be positive with probability 0.999; if he is well, the test will be positive with probability 0.2. A person takes the x-ray test, which gives a positive result. What is the probability that the person is well?

Again, let A_1 be the event "tuberculosis," A_2 the complementary event. Let B_1 be the event "positive." We have

$$P(A_1) = 0.01 \qquad P(A_2) = 0.99$$

$$P(B_1|A_1) = 0.999 \qquad P(B_1|A_2) = 0.2$$

Then

$$P(A_2|B_1) = (0.2)(0.99)/[(0.01)(0.999)+(0.2)(0.99)] =$$

$$0.198/(0.00999+0.198) = 0.95$$

Note that even with a positive test result. the person tested is still more likely to be well than ill. The reason for this is simply that it is considered better for a healthy patient to undergo further examinations than for a sick man to go undetected. The x-ray test is made purposely difficult to pass, i.e., the smallest dark spot is considered a positive result.

PROBLEMS ON BAYES' FORMULA

1. A company has a position open; the probability that an applicant be capable of doing the required work is 0.65. The firm's personnel department devises an aptitude test that a capable applicant will pass with probability 0.8, whereas an incapable applicant will pass with probability 0.4. An applicant passes the test; what, is the probability that he will be able to do the work?

2. Three identical urns are labeled A, B, and C. Urn A contains 5 red and 3 black balls. Urn B contains 6 red and 10 black halls; urn C contains 12 red and 4 black balls. An urn is chosen at random, and then a ball is taken from the urn, also at random. If the ball is red, what is the probability that the urn was urn B?

3. A person is either well (probability 0.9) or ill (probability 0.1). If he is well, his temperature will be normal with probability 0.95, whereas if he is ill, his temperature will be normal with probability 0.35.

(a) If a person's temperature is normal, what is the probability that he is ill?
(b) If the temperature is not normal, what is the probability that he is well?

4. A machine makes bolts to fit specifications. When the; machine is working well (probability 0.8) the bolts fit specifications with probability 0.95; when it is not working well, the bolts fit specifications with probability 0.5. Two bolts taken from the machine's production are both defective. What is the probability that the machine is working well?

5. A contractor must build a road through a piece of ground. The soil is either clay (probability 0.68) or rock. If it is rock, a geological test will give a positive result with probability 0.75. If it is clay, the same test will give a positive result with probability 0.25. Given that the test shows a positive result) what is the probability that the soil is rock?

6. REPETITION OF SIMPLE EXPERIMENTS; THE BINOMIAL DISTRIBUTION

Let us consider a very elementary type of experiment: one with only two outcomes. These outcomes are generally called *success* (S) and *failure* (F). Such an experiment is called a *simple experiment*. It may be described entirely in terms of the single number p, where

3.6.1 $$p = P(S)$$

Since the probability of success is p, it follows that the probability of failure is $1-p$; we write

3.6.2 $$q = 1-p = P(F)$$

For a single trial of this experiment there is very little else to say. Suppose, however, that the experiment is repeated many times. We will obtain a run of successes and failures, which may look like

SFFSSFFSSFSFFSSS

We are interested, generally, not so much in the exact run of successes and failures, but rather, in the number of successes obtained.

Consider the run just given. The probability of each success is p; that of each failure is q. If we assume (as we shall) that the trials are independent, then the probability of this run is

pqqppqqppqpqqppp

obtained by multiplying the probabilities for each trial. This may, of course, be rewritten as $p^9 q^7$. Note that any other run of nine successes and seven failures would have the same probability. In general, we find that, in n independent trials of a simple experiment, any run that contains k successes and $n-k$ failures has probability $p^k q^{n-k}$. To find the probability of obtaining exactly k successes in n trials, we must find the number of runs having k successes. This number is obtained by noticing that the k successes can appear in any k of the n trials; thus, the number of such runs must be equal to the number of combinations of k elements that can be taken from a set of n elements, i.e. $_nC_k$. Multiplying this number by the probability of each run, we obtain the *binomial probability*

3.6.3 $$B(n,k;p) = {}_nC_k p^k q^{n-k}$$

This expression is known as a Bernoulli coefficient, after Jakob Bernoulli (1654-1705), who analyzed the problem. A sequence of independent trials (of a simple experiment) is also known as a sequence of Bernoulli trials.

III.6.1. Example. A fair coin is tossed ten times. What is the probability of obtaining exactly seven heads?

The coin is fair, so $p = 1/2$. According to formula (3.6.3), the probability of exactly seven heads is

$$B(10,7;1/2) = {}_{10}C_7 (1/2)^7 (1/2)^3 = 120/1024$$

III.6.2. Example. A fair die is tossed five times. What is the probability of exactly three aces?

In this case, $p = 1/6$. Hence

$$B(5,3;1/6) = {}_5C_3 (1/6)^3 (5/6)^2 = 250/7776$$

III.6.3. Example. A machine produces bolts to meet certain specifications; 90 per cent of the bolts produced meet those specifications. A sample of five bolts is taken from the machine's production. What is the probability that two or more of these fail to meet the specifications?

In this problem, we are looking for the event "at least two fail to meet specifications," or, more concisely, "at least two failures." The formula given, however, only serves to get the probability of exactly k successes (or failures). Thus we divide the event into the disjoint events "2 failures," "3 failures," "4 failures," and "5 failures." We have $n = 5$ and $p = 0.9$, and so the relevant probabilities are

$$P(2 \text{ failures}) = B(5,3;0.9) = {}_5C_3 (0.9)^3 (0.1)^2 = 0.0729$$

$$P(3 \text{ failures}) = B(5,2;0.9) = {}_5C_2 (0.9)^2 (0.1)^3 = 0.0081$$

$$P(4 \text{ failures}) = B(5,1;0.9) = {}_5C_1 (0.9)(0.1)^4 = 0.00045$$

$$P(5 \text{ failures}) = B(5,0;0.9) = {}_5C_0 \, (0.1)^5 = 0.00001$$

and so

$$P(\text{at least 2 failures}) = 0.0729 + 0.0081 + 0.00045 + 0.00001 = 0.08146.$$

This same problem could have been solved by considering the complementary event, "at most 1 failure." This can be written as the union of the two events, "one failure," and " no failures." We have

$$P(0 \text{ failures}) = B(5,5;0.9) = {}_5C_5 \, (0.9)^5 = 0.59049$$

$$P(1 \text{ failures}) = B(5,4;0.9) = {}_5C_4 \, (0.9)^4 \, (0.1) = 0.32805$$

so that

$$P(\text{at most one failure}) = 0.59049 + 0.32805 = 0.91854.$$

Since the events "at most one failure" and "at least two failures" are complementary, we have

$$P(\text{at least two failures}) = 1 - 0.91854 = 0.08146$$

Note that, as expected, we obtain the same answer both ways.

III.6.4. Example. In a large city, 60 per cent of the heads of households are Democrats. A poll taker visits eight houses, and asks the party affiliation of the head of household. What is the probability that at least six are Democrats?

Because the city is large, it may be assumed that the probabilities remain equal at each trial; i.e., the trials are independent. The probability of *at least six Democrats*, then, is

$$B(8,6;0.6) + B(8,7;0.6) + B(8,8;0.6)$$

Now,

$$B(8,6;0.6) = {}_8C_6 \, (0.6)^6 (0.4)^2 = 0.209$$

$$B(8,7;0.6) = {}_8C_7 \, (0.6)^7 (0.4) = 0.089$$

so that
$$B(8,8;0.6) = {}_8C_8 (0.6)^8 = 0.017$$

$$P(\text{at least six Democrats}) = 0.315$$

7. DRAWINGS WITH AND WITHOUT REPLACEMENT

Consider, next, the following problem:
An urn contains 10 red and 4 black balls. Three balls are drawn from the urn. What is the probability that exactly two of the balls will be red?

In a problem such as this, in which elements are taken from a finite set (drawings from a finite population), there are two possible and distinct methods of carrying out the experiment. The experimenter may, if he wishes, take a ball from the urn, record its color, and *replace it* before drawing the next ball. This method is known as *drawing with replacement*. Alternatively, the experimenter may draw the three balls, simultaneously or one at a time, without replacing them. This method is known as *drawing without replacement*. It is clear that the two methods will give different results: for instance, if the urn contains only two balls, it will be impossible even to make three drawings without replacement, whereas any number of drawings is possible with replacement.

Suppose that the drawings are made with replacement. In this case, it is not difficult to see that, at each drawing, the urn contains the original number of red and black balls. The probability of obtaining a red ball will be the same at each drawing, regardless of what happened on the previous draws; i.e., the drawings are independent. It follows that such drawings are Bernoulli trials. Thus, for drawings with replacement, the solution to the problem is as follows. Since 10 of the 14 balls are red, we have $p = 5/7$. Then the probability of 2 red balls is

$$B(3,2;5/7) = {}_3C_2 (5/7)^2(2/7) = 150/343$$

Consider nest the case of drawings without replacement. For this method, the drawings are no longer independent and so they cannot be treated as Bernoulli trials. A new analysis of the problem is therefore necessary.

We analyze the problem of drawings without replacement by assuming that all the possible samples of three balls that can be drawn from the urn have the same probability. It is then merely a question of seeing how many such samples there are, and how many of those belong to the desired event (i.e., contain two red and one black ball). The desired probability is then the quotient of these two numbers.

In fact, the number of possible samples is equal to the number of combinations of three elements from a set of fourteen, i.e. to the binomial coefficient $_{14}C_3$. In turn, the number of samples containing 2 red and 1 black balls is obtained by multiplying the coefficients $_{10}C_2$ and $_4C_1$. There are then $_{14}C_3$ possible samples, of which $_{10}C_2$ $_4C_1$ belong to our event. Thus,

$$P(\text{exactly two red balls}) = {}_{10}C_2 \, {}_4C_1 / {}_{14}C_3 = 36/91$$

Note that the probability in this case is slightly less than in the case of drawings with replacement.

In general, we may postulate an urn containing m balls, of which m_1 are red and $m_2 = m - m_1$ are black. A sample of n balls is taken (without replacement) from the urn; we want the probability that exactly k of them are red.

There are, all told, $_mC_n$ combinations of n elements from a set of m elements. This is the number of possible samples. Of these, we consider those tyhat have exactly k red and $n-k$ black balls. There are $_{m1}C_k$ combinations of k red balls, and $_{m2}C_{n-k}$ combinations of $n-k$ black balls. The product of these two numbers will be the number of samples in the event. Thus,

3.7.1 $\qquad P(\text{exactly } k \text{ red balls}) = {}_{m1}C_k \, {}_{m2}C_{n-k} / {}_mC_n$

III.7.1. Example. A poker hand consists of 5 cards, taken at random from a deck of 52 (which is divided into 4 suits of 13 cards each). What is the probability that all cards in the hand will be of the same suit (a hand know as a flush)?

We can best solve this problem by considering the probability that all five cards in the hand are, say, spades. The probability of a flush will be four times this, since the flush can be in any one of the four suits.

It is easy to see that this is a problem of sampling without replacement. We will have $m = 52$, $m_1 = 13$ (the number of spades), and $m_2 = 39$. Also, $n = k = 5$.

$$P(\text{all 5 are spades}) = {}_{13}C_5 \, {}_{39}C_0 / {}_{52}C_5 = 33/66640$$

and therefore

$$P(\text{flush}) = 4 \, P(\text{flush in spades}) = 33/16660$$

The probability of a flush is slightly less than one in 500. (This probability is not exactly the same as that given in poker primers because we include straight flushes among the flushes.)

III.7.2 Example. A company tests a lot of 50 flash-bulbs by taking a sample of 10. If the sample contains 2 or more duds, the lot is rejected. Given that the lot contains 13 duds, what is the probability that it will be rejected?

As in Example III.6.3, we shall do this by considering the complementary event, "at most one dud." This, in turn, is divided into the two events, "no duds," and "exactly one dud." Now,

$$P(\text{no duds}) = {}_{13}C_0 \, {}_{37}C_{10} / {}_{50}C_{10} = 0.033$$

$$P(\text{one dud}) = {}_{13}C_1 \, {}_{37}C_9 / {}_{50}C_{10} = 0.153$$

and so the probability that the lot will be rejected is

$$1 - 0.033 - 0.153 = 0.814$$

III.7.3. Example. A small town has 20 houses. In 12 of these, the head of household is a Democrat. A poll taker visits 8 houses. What is the probability that in at least 6, the head of household be a Democrat?

As opposed to Example III.6.4, this is a case of sampling without replacement. We have

$$P(6 \text{ democrats}) = {}_{12}C_6 \, {}_8C_2 / {}_{20}C_8 = 0.205$$
$$P(7 \text{ democrats}) = {}_{12}C_7 \, {}_8C_1 / {}_{20}C_8 = 0.050$$
$$P(8 \text{ democrats}) = {}_{12}C_8 \, {}_8C_0 / {}_{20}C_8 = 0.004$$

so that

$$P(\text{at lest 6 democrats}) = 0.259$$

PROBLEMS ON DRAWINGS WITH AND WITHOUT REPLACEMENT

1. A fair coin is tossed 12 times. What is the probability of obtaining:
   (a) Exactly 6 heads?
   (b) At least eight heads?
   (c) Between five and seven heads?

2. A fair die is tossed five times. What is the probability of obtaining:

(a) Exactly two aces?
(b) At least three aces?

3. A multiple-choice test consists of eight questions, each having three possible answers. What should the passing mark be to ensure that someone who guesses at the answers will have a probability smaller than 0.05 of passing?

4. A machine produces pieces to meet specifications; the probability that a given piece be suitable is 0.95. A sample of 10 pieces is taken. What is the probability that 2 or more be defective?

5. Two baseball teams meet in the World Series; team A, the stronger team, has probability 0.6 of winning in each game. What is probability that team A will win the series (i.e., win at least four out of seven games)?

6. A package of flash bulbs contains 16 good and 4 defective bulbs. A sample of 5 bulbs is taken. What is the probability that at least three of the bulbs will be good?

7. An urn contains 15 red and 10 black balls. Three balls are taken from the urn; what is the probability that 2 of these be red?

8. A lot of 16 tires contains 13 good and 3 defective tires. A sample of 6 is taken. What is the probability that it contains at least 4 good tires?

9. Lots of 20 elements are tested in the following manner. A sample of size 4 is taken, and the number of defective elements in the sample is called x. If $x = 0$, the lot is accepted, if $x \geq 3$, it is rejected. If $x = 1$ or 2, a further sample of size 3 is taken and the number of defectives in this lot is called y. Then the lot is accepted if $x + y \leq 3$, and rejected if $x + y \geq 4$. What is the probability that a lot will be rejected if:
(a) It contains 15 good and 5 defective elements?
(b) It contains 12 good and 8 defective?
(c) It contains 17 good and 3 defective?

10. A simple experiment, with probability of success p, is repeated until m successes are obtained. Show that the probability of obtaining the mth success on the kth trial is given by the *negative binomial distribution*

$$P_m(k) = {}_{k-1}C_{m-1}\, p^m\, (1-p)^{k-m}$$

8. RANDOM VARIABLES

Very often there is a numerical quantity associated with each outcome of an experiment. It may represent money, as in the case of a gamble, or it may represent some other measurable quantity - the strength of a rope, the number of games won by a baseball team, or the life of a light bulb. Such a numerical quantity is known as a *random variable*. We give a precise definition.

III.8.1. Definition. A *random variable* is a function whose domain is the set of outcomes of an experiment and whose range is a subset of the real numbers.

We shall generally denote a random variable by a capital letter X, Y, and so on. The values of this random variable will be denoted by lower-case letters (x, y, and so on).

Since, by its very nature, a random variable takes its values according to a randomization scheme, it becomes natural to look for the probability of each of these values. These probabilities make up what is known as the *distribution* of the random variable.

Associated with the random variable X, there are two functions that describe its distribution. The first is the *probability* function $f(x)$, defined by

3.8.1 $$f(x) = P(X = x)$$

The second is the *cumulative distribution* function, $F(x)$, defined by

3.8.2 $$F(x) = P(X \leq x)$$

Thus f gives tile probabilities of each of the several possible values of X, while F gives the probability that X will be smaller than or equal to a given number. We shall, almost invariably, deal with the probability function, f, although the cumulative distribution function will be of use in some occasions.

III.8.2. Example. Let X be the number of spots showing on one toss of a fair die. Then X has the probability function

$$f(x) = 1/6 \quad \text{for } x = 1,2,\dots,6$$

0 otherwise

FIGURE III.8.1. Probabilities for the toss of one die.

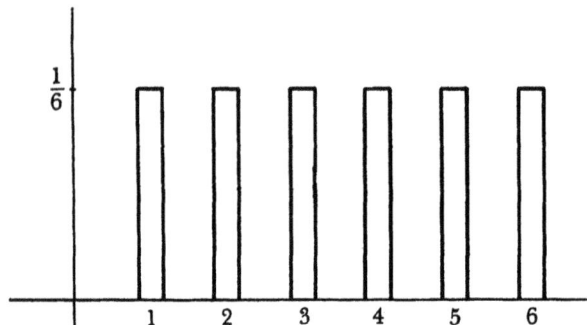

and cumulative distribution function

$$F(x) = \begin{cases} 0 & \text{for} \quad x < 1 \\ 1/6 & \text{for} \quad 1 \leq x < 2 \\ 1/3 & \text{for} \quad 2 \leq x < 3 \\ 1/2 & \text{for} \quad 3 \leq x < 4 \\ 2/3 & \text{for} \quad 4 \leq x < 5 \\ 5/6 & \text{for} \quad 5 \leq x < 6 \\ 1 & \text{for} \quad x \geq 6 \end{cases}$$

See Figures III.8.1 and III.8.2 for these two functions.

III.8.3. Example. Let Y be the number of spots appearing on two independent tosses of a fair die. The distribution of Y may be found by studying the table in Example III.2.7; it is

$$f(2) = 1/36$$
$$f(3) = 2/36$$
$$f(4) = 3/36$$
$$f(5) = 4/36$$
$$f(6) = 5/36$$
$$f(7) = 6/36$$
$$f(8) = 5/36$$
$$f(9) = 4/36$$

$$f(10) = 3/36$$
$$f(11) = 2/36$$
$$f(12) = 1/36$$

FIGURE III.8.2. Cumulative distribution for the toss of one die.

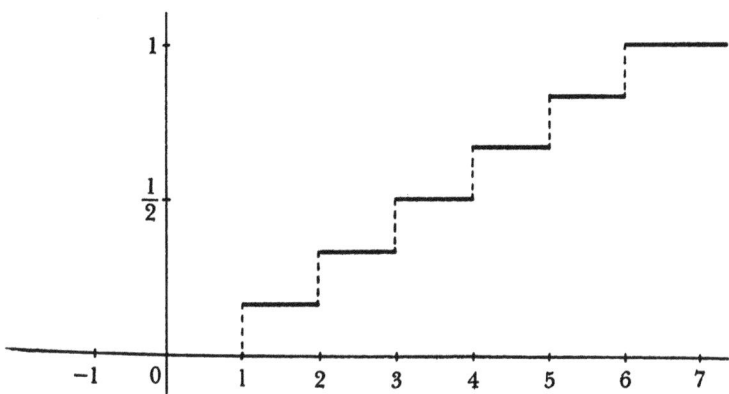

and $f(y) = 0$ for all other y.

FIGURE III.8.3. Probabilities for the toss of two dice.

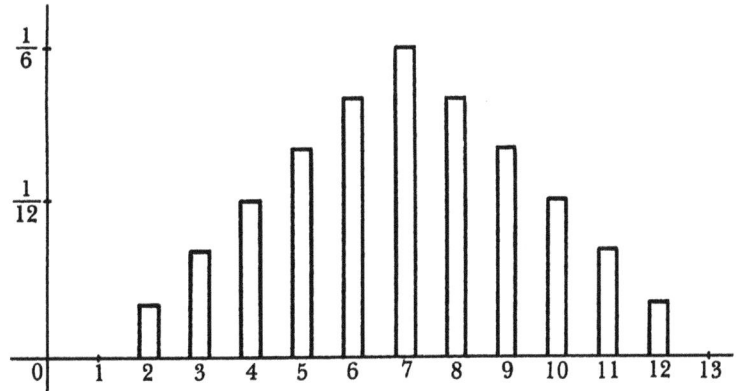

III.8.4. Example. Let X be the number of heads obtained in 10 independent tosses of a fair coin. Its distribution is obtained by using formula (6.4.3) with $n = 10$, $p = 1/2$.

$$f(0) = {}_{10}C_0 (1/2)^{10} = 1/1024$$

$$f(1) = {}_{10}C_1 (1/2)(1/2)^9 = 10/1024$$
$$f(2) = {}_{10}C_2 (1/2)^2 (1/2)^8 = 45/1024$$
$$f(3) = {}_{10}C_3 (1/2)^3 (1/2)^7 = 120/1024$$
$$f(4) = {}_{10}C_4 (1/2)^4 (1/2)^6 = 210/1024$$
$$f(5) = {}_{10}C_5 (1/2)^5 (1/2)^5 = 252/1024$$
$$f(6) = {}_{10}C_6 (1/2)^6 (1/2)^4 = 210/1024$$
$$f(7) = {}_{10}C_7 (1/2)^7 (1/2)^3 = 120/1024$$
$$f(8) = {}_{10}C_8 (1/2)^8 (1/2)^2 = 45/1024$$
$$f(9) = {}_{10}C_9 (1/2)^9 (1/2) = 10/1024$$
$$f(10) = {}_{10}C_{10} (1/2)^{10} = 1/1024$$

In Example III.8.4, note that while 5 is the most probable value of the random variable, its probability is still less than 1/4. On the other hand, there is approximately a 0.66 probability that the variable will be between 4 and 6.

FIGURE III.8.4. Binomial distribution with n = 10, p - 1/2.

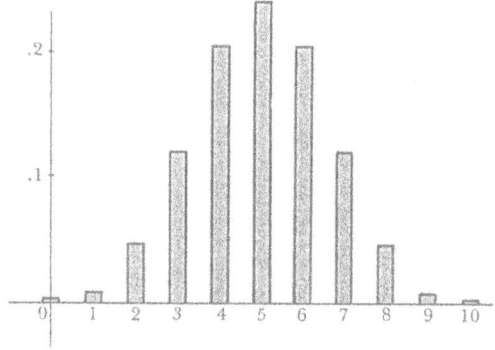

If X and Y are random variables that depend on the outcome of the same experiment, we consider their *joint distribution*, given by the function f:

3.8.3 $$f(x,y) = P(X = x, Y = y)$$

Thus in the experiment of tossing a fair die twice, we may let X and Y be the number of spots showing on the first and second tosses of the die, respectively. Then the joint probability of X and Y is given by

$$f(x,y) = 1/36 \text{ for } x, y = 1, 2, ..., 6$$

0   otherwise

The *absolute, or marginal, distribution* of X may be obtained from the joint distribution of X and Y; it is given by

3.8.4 $$g(x) = \sum_j f(x,y_j)$$

where the summation is taken over all the possible values, y_j, of the variable Y. Similarly, the absolute distribution of Y is given by

3.8.5 $$h(y) = \sum_i f(x_i,y)$$

where the summation is taken over all the possible values, x_i, of X. Closely connected to these are the *conditional* distributions of X and Y. For a possible value, y_j, of Y, we define the conditional distribution of X to be

3.8.6 $$g(x|y_j) = f(x,y_j)/h(y_j)$$

Similarly, for one of the possible values, x_i, of X, the conditional distribution of Y is

3.8.7 $$h(y|x_i) = f(x_i,y)/g(x_i)$$

III.8.5. Example. Let X be the number of heads obtained in three tosses of a fair coin, and let Y be either 1 or 0, depending on whether the first toss is heads or tails. Then the joint distribution of X and Y is given by Table III.8.1.

TABLE III.8.1.

Y	X				h(y)
	0	1	2	3	
0	1/8	1/4	1/8	0	1/2
1	0	1/8	1/4	1/8	1/2
g(x)	1/8	3/8	3/8	1/8	

The marginal row and marginal column give the absolute distributions of X and Y, respectively. The conditional distributions are easily obtained; we might give them in the form of two tables. The first table, III.8.2, gives the conditional distribution, $g(x|y_j)$; the second, Table III.8.3, gives $h(y|x_i)$:

TABLE III.8.2.

Y	X				
	0	1	2	3	
0	1/4	1/2	1/4	0	$g(x|y_j)$
1	0	1/8	1/4	1/8	

TABLE III.8.3.

Y	X				
	0	1	2	3	
0	1	2/3	1/3	0	$h(y|x_i)$
1	0	1/3	2/3	1	

As may be seen in Example III.8.5, the conditional distributions of a random variable may be quite different from its absolute distribution. In some cases, however, these are all equal: the random variables are then independent. We give a more formal definition.

III.8.6.Definition. Two random variables X and Y are *independent* if, for every pair (x,y) of values of the two variables.

$$f(x,y) = g(x)\, h(y)$$

An equivalent definition of independence would be to say that two variables are independent if their joint probability function can be factored into two functions, each of which depends on only one of the two variables.

For instance, the random variables X, Y of Example III.8.4 are independent. On the other hand, the variables of Example III.8.5 are not independent.

III.8.7 Example. Consider the date of Example III.2.7. We can let X be 0 or 1 according to whether the family is Republican or Democratic, and Y be 0 if the family is not headed by a woman, 1 if it is. Then the joint distribution of (X,Y) is shown in Table III.8.4.

TABLE III.8.4

Y	X		h(y)
	0	1	
0	21/39	11/39	32/39
1	2/39	5/39	7/39
g(x)	23/39	16/39	

The two marginal distributions, $g(x)$ and $h(y)$, are given are given at the bottom ands to the right, respectively, of the joint distribution table.

The conditional distributions are shown in Tables III.8.5 and III.8.6.

TABLE III.8.5

Y	X		
	0	1	
0	21/32	11/32	$g(x\|y_j)$
1	2/7	5/7	

TABLE III.8.6

Y	X		
	0	1	
0	21/23	11/16	$h(y\|x_i)$
1	2/23	5/16	

Note that X and Y are *not* independent.

9. EXPECTED VALUES. MEANS AND VARIANCES

In General, if a random variable has any practical interpretation (if, for instance, it represents money or some measurable service or commodity) it becomes natural to look, in some way, for an *average* value of the variable. An example can explain this.

Consider the game of roulette, as played in European casinos. The roulette wheel has 37 numbers ranging from 0 to 36. There are many types of bet possible, but the simplest is a bet on a single number. If the ball lands on the chosen number, the player gets back his bet, plus 35 times his bet. If it lands on any other number, he loses his bet.

Assume the player bets $1 on a single number. Let X be the amount won by the player; then X can have the values −1 or +35. The probabilities of these two values can be obtained by assuming that the roulette wheel is perfectly symmetric, so that the ball has the same probability (i.e., 1/37) of landing on any of the numbers. Then

$$f(+35) = 1/37$$
$$f(-1) = 36/37$$

Suppose, next, that the player makes this bet, not once, but very many times. If the total number of bets made is n, then, by the frequency definition of probability, he should win approximately $n/37$ of the times, and lose approximately $36n/37$ times. Since each win is $35, and each loss is $1, his total winnings should be (approximately) $35n/37$, his total losses should be $36n/37$, and so his net winnings would be $-n/37$.

Since we are interested in average (per bet) winnings, we divide these net winnings by n (the number of bets), and find an average of $-1/37$ dollars per bet: this is what we might *expect* over a long string of bets. (Of course, the bettor might have a lucky streak and win, or an unlucky streak and lose much more than that, but this is what we expect.) Thus, this $-1/37$ is the *expected value* of X.

More generally, suppose that the random variable X has the possible values x_1, x_2, ..., x_m with probabilities $f(x_1)$, $f(x_2)$, ... $f(x_m)$ respectively. If the experiment is repeated a large number, n, of times, then X should take on the value x_i approximately $nf(x_i)$ times. If there is any point to adding the values of X obtained, we find that the sum of the values of X, for the (approximately) $nf(x_i)$ times that it has the value x_i, is $nx_i f(x_i)$. Summing these expressions for all the possible values of X, we obtain

$$\sum_{i=1}^{m} nx_i f(x_i) = n \sum_{i=1}^{m} x_i f(x_i)$$

To get an average value, we now divide this by the number of trials, and obtain the somewhat simpler form

3.9.1
$$E[X] = \sum_{i=1}^{m} x_i f(x_i)$$

This expression is the *expected value*, or *expectation*, of the random variable X.

III.9.1. Example. Let X be the number of spots showing after a toss of a fair die. Its expectation can be calculated from Table III.9.1.

TABLE III.9.1

x_i	$f(x_i)$	$x_i f(x_i)$
1	1/6	1/6
2	1/6	2/6
3	1/6	3/6
4	1/6	4/6
5	1/6	5/6
6	1/6	6/6

Adding the entries in the last column of this table, we obtain the expectation

$$E[X] = 7/2$$

III.9.2. Example. Let X be the number of heads obtained in 10 tosses of a fair coin. We form Table III.9.2.

TABLE III.9.2

x_i	$f(x_i)$	$x_i f(x_i)$
0	1/1024	0
1	10/1024	10/1024
2	45/1024	90/1024
3	120/1024	360/1024
4	210/1024	840/1024
5	252/1024	1260/1024
6	210/1024	1260/1024
7	120/1024	840/1024
8	45/1024	360/1024
9	10/1024	90/1024
10	1/1024	10/1024

Once, again, we add the entries in the last column, to obtain

$$E[X] = 5$$

If X is a random variable, and U is a function of X, say $U = \varphi(X)$, then U is also a random variable. We may obtain the expected value of U by computing its distribution (which can be done if we know the distribution of X). This is not generally necessary, however, as it may be seen that

3.9.3 $$E[\varphi(X)] = \sum_{i=1}^{m} \varphi(x_i) f(x_i)$$

III.9.3. Example. Let X be as in Example III.9.1, and let $U = X^2$. We compute its expectation from Table III.9.3.

TABLE III.9.3

x_i	$f(x_i)$	$x_i^2 f(x_i)$
1	1/6	1/6
2	1/6	4/6
3	1/6	9/6
4	1/6	16/6
5	1/6	25/6
6	1/6	36/6

Adding the entries in the last column, we obtain

$$E[U] = 91/6$$

Two of the most important descriptive properties of a random variable are its *mean* and its *variance*. These are defined in terms of expectations.

III.9.4. Definition. Let X be a random variable. Then the *mean*, μ_X, and the *variance*, σ_X^2, of X, are defined as

3.9.3 $$\mu_X = E[X]$$

3.9.4 $$\sigma_X^2 = E[(X-\mu_X)^2]$$

The mean is the same as the expected value of X. The variance, which is always non-negative, is a measure of the *dispersion* of X: If σ_X^2 is small, then X will usually be close to μ_X, whereas if σ_X^2 is large, then X will often differ considerably from its mean. The square root of the variance, σ_X, is known as the *standard deviation*.

10. RULES FOR COMPUTING THE MEAN AND VARIANCE

Computation of the mean and variance of a random variable is often simplified by the following theorem:

III.10.1. Theorem. Let X and Y be random variables, and α a constant. Then

3.10.1 $\quad E[\alpha] = \alpha$
3.10.2 $\quad E[\alpha X] = \alpha E[X]$
3.10.3 $\quad E[X+Y] = E[X] + E[Y].$

Moreover, if X and Y are independent, then

3.10.4 $\quad E[XY] = E[X]E[Y]$

Proof. If α is a constant (not a variable) then it will have the value α with probability 1, and no other values. This proves (3.10.1).

To prove (3.10.2), note that

$$E[\alpha X] = \sum \alpha x_i f(x_i) = \alpha \sum x_i f(x_i) = \alpha E[X]$$

To prove (3.10.3), let X and Y have the joint distribution f, and let g and h be the absolute distributions of X and Y respectively. Then

$$\begin{aligned} E[X+Y] &= \sum_{i,j}(x_i+y_j)f(x_i,y_j) \\ &= \sum_{i,j} x_i f(x_i,y_j) + \sum_{i,j} y_j f(x_i,y_j) \\ &= \sum_i \{x_i \sum_j f(x_i,y_j)\} + \sum_j \{y_j \sum_i f(x_i,y_j)\} \\ &= \sum_i x_i g(x_i) + \sum_j y_j h(y_j) \\ &= E[X] + E[Y]. \end{aligned}$$

Finally, suppose X and Y are independent. This means $f(x,y) = g(x)h(y)$. Then

$$\begin{aligned} E[XY] &= \sum_{i,j} x_i y_j f(x_i,y_j) \\ &= \sum_{i,j} (x_i g(x_i))(y_j h(y_j)) \\ &= \{\sum_i x_i g(x_i)\} \{\sum_j y_j h(y_j)\} \\ &= E[X]E[Y] \end{aligned}$$

III.10.2. Corollary. If X has mean μ and variance σ^2, then

3.10.5
$$\sigma^2 = E[X^2] - \mu^2$$

Proof. We have

$$\sigma_X^2 = E[(X-\mu)^2] = E[X^2 - 2\mu X + \mu^2]$$
$$= E[X^2] - 2\mu E[X] + \mu^2$$
$$= E[X^2] - 2\mu^2 + \mu^2$$
$$= E[X^2] - \mu^2$$

III.10.3. Corollary. If X and Y are independent random variables with variances σ_X^2 and σ_Y^2 respectively, then the variance of $X + Y$ is

3.10.6
$$\sigma_{X+Y}^2 = \sigma_X^2 + \sigma_Y^2$$

Proof. We have

$$E[(X+Y)^2] = E[X^2 + 2XY + Y^2]$$
$$= E[X^2] + 2E[XY] + E[Y^2]$$

and

$$\mu_{X+Y}^2 = (\mu_X + \mu_Y)^2 = \mu_X^2 + 2\mu_X\mu_Y + \mu_Y^2$$

Now, by independence, $E[XY] = \mu_X\mu_Y$, and so

$$\sigma_{X+Y}^2 = E[(X+Y)^2] - \mu_{X+Y}^2$$
$$= E[X^2] - \mu_X^2 + E[Y^2] - \mu_Y^2$$
$$= \sigma_X^2 + \sigma_Y^2$$

III.10.4. Corollary. Let X be a random variable with mean μ_X and variance σ_X^2, and let $Y = \alpha X + \beta$. Then Y has mean

3.10.7 $$\mu_Y = \alpha\mu_X + \beta$$

and variance

3.10.8 $$\sigma_Y^2 = \alpha^2\sigma_X^2.$$

We will omit the proof of this last corollary.

11. TWO IMPORTANT THEOREMS

The importance of the mean and standard deviation of a random variable is given by the following two theorems.

III.11.1. Theorem (Chebyshev's Inequality). Let X have mean μ and standard deviation σ. Then, for any $t > 0$,

3.11.1 $$P(|X-\mu| \geq t\sigma) \leq 1/t^2$$

Proof. Assume that the event

$$|X-\mu| \geq t\sigma$$

or, equivalently,

3.11.2 $$(X-\mu)^2 \geq t^2\sigma^2$$

has probability p. We may partition the values of X into two sets: those for which (3.11.2) holds, and those for which it does not. We may write

$$\sigma^2 = \Sigma_i (x_i - \mu)^2 f(x_i)$$
$$= \Sigma_i' (x_i - \mu)^2 f(x_i) + \Sigma_i'' (x_i - \mu)^2 f(x_i)$$

where the first sum corresponds to those values x_i for which (3.11.2) holds, and the second, to the other values of X. Now, the second sum is non-negative, and so

$$\sigma^2 \geq \Sigma_i' (x_i - \mu)^2 f(x_i)$$

and, as (3.11.2) holds for this sum,

$$\sigma^2 \geq \Sigma_i' \, t^2 \sigma^2 f(x_i)$$
$$= t^2 \sigma^2 \Sigma_i' \, f(x_i).$$

This last sum is precisely the probability of (3.11.2). Hence

$$\sigma^2 \geq t^2 \sigma^2 p$$

and so

$$p \leq 1/t^2$$

Chebyshev's Inequality states that a large deviation from the mean can never be too probable; this is given in terms of the standard deviation.

III.11.2. Theorem. (The Law of Large Numbers). Let X be a random variable with mean μ. Let $X_1, X_2, \ldots, X_n$ be n independent repetitions of the variable X, and let

$$M_n = 1/n \sum_{i=1}^{n} X_i$$

Then, for any $\varepsilon > 0$, and any $\delta > 0$, there exists n such that

3.11.3
$$P(|M_n - \mu| > \varepsilon) < \delta$$

Proof. Assume X has variance σ^2. Then

$$Y_n = \sum_{i=1}^{n} X_i$$

will have variance $n\sigma^2$, and so $M_n = Y_n/n$ will have variance σ^2/n. Applying Chebyshev's inequality, we see that

$$P(|M_n - \mu| > \varepsilon) \leq \sigma^2/n\varepsilon^2$$

And it is easy to see that, as n increases, the right side of this inequality approaches 0.

The Law of Large Numbers states that the idea of an expected value is really a valid idea: if the experiment is repeated sufficiently many times, it becomes almost certain that that M_n, the average observed value of X, will be very close to μ. Note that this does not say that any particular value of M will be very probable, only that those values which are close to μ will be much more probable than those which are not. For example, in 100 tosses of a fair coin, the most probable number of heads is 50. Yet the probability of having exactly 50 heads is only 0.08. This is not large but, on the other hand, the probability of having exactly 40 heads is only 0.01, while that of having exactly 35 heads is 0.001. The probability of having between 40 and 60 heads (i.e. of having $0.4 \leq M \leq 0.6$) will, however, be about 0.96.

III.11.3. Example. Find the mean and variance of the number of successes, X, in n independent trials of a simple experiment with probability of success p.

Assume first that $n = 1$. In this case, X can have only the values 0 and 1, with probabilities q and p respectively. Then

$$E[X] = 0 \cdot q + 1 \cdot p = p$$

and

$$E[X^2] = 0^2 \cdot q + 1^2 \cdot p = p$$

It follows that X has mean $\mu = p$, and variance

$$\sigma^2 = E[X^2] - \mu^2 = p - p^2 = pq.$$

For general values of n, we define the random variable Y_i, which is 0 if there is a failure on the ith trial, and 1 if there is a success in the ith trial. Then it is not difficult to see that

$$X = Y_1 + Y_2 + \ldots + Y_n$$

Now, the Y_i are independent random variables, each having mean p and variance pq. It follows that the sum of n of them must have mean and variance

3.11.9 $$\mu = np, \qquad \sigma^2 = npq.$$

(Remember that the standard deviation, which is frequently of interest, is the square root of the variance, and thus proportional only to the *square root* of *n*.)

III.11.4. Example. Let X be the (sum of the) number of spots showing on n independent tosses of a die. Find the mean and variance of X.

In case n = 1, we saw above that the $E[X] = 7/2$ and the $E[X^2] = 91/6$. Thus

$$\mu = 7/2, \quad \sigma^2 = 91/6 - (7/2)^2 = 35/12$$

For n independent tosses, we simply add the means and variances of the n tosses, i.e. multiply by n. Thus

$$\mu = 7n/2, \quad \sigma^2 = 35n/12$$

III.11.5. Example. A gambler bets \$1 on a simple experiment which has probability p of success. If the outcome is a success, he wins \$1 and stops playing. If the outcome is a failure, he loses \$1 and bets \$2 on the next trial of the experiment. He continues in this way, doubling his bet after each failure, until he obtains a success. He stops betting if he either obtains a success or loses n times in a row (an event which he considers extremely unlikely). What is the expected value of his winnings?

The gambler's bet on the kth trial (assuming he is still betting) is 2^{k-1} dollars. If he loses on each of the first k trials, he will have lost

$$1 + 2 + 2^2 + 2^3 + \ldots + 2^{k-1} = 2^k - 1$$

dollars before betting on the kth trial. The bet will then be 2^k dollars, so that, if he wins, he will win back all his losses plus an additional dollar. If, however, he loses n consecutive times, then he will wind up losing a total of $2^n - 1$ dollars. Letting X represent his winnings, we see that X can have the two values $+1$ and $-(2^n - 1)$. The probability of losing n consecutive times is q^n and so

$$E[X] = 1(1-q^n) + (1-2^n)q^n$$

$$= 1 - 2^n q^n$$

$$= 1 - (2q)^n$$

Now, if $q > \frac{1}{2}$, then $2q > 1$, and so $(2q)^n$ increases without bound as n increases. For large values of n, therefore, the variable X has an expected value that is negative and very large. This is true even though large values of n mean that the probability q^n of losing is very small. We conclude that this system of betting (known as a *martingale*) is not very good.

III.11.6. Example. In a certain large city, 55 per cent of the population are Democrats. A pollster takes a sample of n people, asking their party affiliation. How large should n be, in order that there be a probability of at least 0.95 that between 54 per cent and 56 per cent of those sampled be Democrats?

Since the city is large, the trials may be considered independent. In a sample of size n, the number X of Democrats will be a binomial random variable with mean

$$\mu = 0.55n$$

and variance

$$\sigma^2 = (0.55)(0.45)n = 0.2475n.$$

The event "between 54 per cent and 56 per cent are Democrats" can be expressed as

$$|X - \mu|/n < 0.01$$

or, equivalently,

$$|X - \mu| < 0.01n$$

We apply Tchebyshev's inequality, with $t\sigma = 0.01\,n$, or

$$t^2 = 0.0001n^2/0.2475n = n/2475.$$

The probability of the desired event is at least $1 - 1/t^2$; thus we therefore want $1/t^2 \leq 0.05$. It is sufficient to take

$$2475/n \leq 0.05$$

or, equivalently, $n \geq 49{,}500$. A sample of size 49,500 will be adequate for our purpose.

In reality, this value of n is unnecessarily large. It may be shown that $n = 9604$ would be sufficient. This is because Chebyshev's inequality uses only the mean and variance of the distribution, and does not take special properties of the distribution (namely, that it is a binomial distribution) into account.

Theorem III.11.2 (The Law of Large Numbers) nearly guarantees that, in the long run, the average value of a random variable will approach its expectation. It becomes natural, then, *if an experiment can be repeated many times*, to take the course of action which will maximize the decision-maker's *expected* profit. This is of importance when the profit (or cost) depends on both the decision-maker's action and the outcome of an experiment.

III.11.7. Example. An automobile company produces a deLuxe model which costs $40,000 to produce and sells for $85,000. The demand for this model is uncertain; the company estimates (from past experience) that as many as four cars may be sold; the probability that k cars will be sold is given in table III.11.1.

TABLE III.11.1

k	$f(k)$
0	0.1
1	0.4
2	0.3
3	0.15
4	0.05

How many cars should the company produce to maximize its (expected) profits?

Let us assume that the company produces n cars, and the demand is for k cars. Assuming a profit of $45,000 for each car sold, and a loss of $40,000 for each car produced but not sold, we obtain the table of profits (Table III.11.2).

TABLE III.11.2

k	n					f(k)
	0	1	2	3	4	
0	0	-40	-80	-120	-160	0.1
1	0	45	5	-35	-75	0.4
2	0	45	90	50	10	0.3
3	0	45	90	135	95	0.15
4	0	45	90	135	180	0.05
E[X]	0	36.5	39	16	20.75	

The right-hand column gives, once again, the probabilities of each value of k. The bottom row gives the expected profits (in thousands of dollars) for each choice of n. We see that this expectation gives a maximum for $n = 2$, and conclude that the company should produce two of these cars every year.

In dealing with expectations, it is assumed (at least tacitly) that the experiment can be repeated many times. Otherwise the law of large numbers has no chance to work. It follows that, when the experiment cannot be repeated, the criterion of maximizing the expected value cannot be applied indiscriminately. Consider, for example, a person who is given the chance of investing $50,000 in a new company. The company has, let us say, a probability 0.5 of succeeding, in which case the investor will receive $200,000 in profits. On the other hand, the company also has probability 0.5 of failure, in which case the investor will lose his $50,000. It is easy to see that the expected profit from this investment is

$$\tfrac{1}{2}(200{,}000) + \tfrac{1}{2}(-50{,}000) = 75{,}000$$

so that, to maximize his expected profits, the investor might take this gamble rather than make some other investment which would give him perhaps a certain profit of $60,000. But is this reasonable?

If the investor is not a wealthy man, he might be unwilling to risk $50,000: he could, quite reasonably, feel that it is more money than he is prepared to lose and that, moreover, he will have no chance to recoup his losses if the company fails (he will then be broke). A certain profit of $60,000 might seem better to him than this risky investment. On the other hand, if the investor is very rich, the gamble might be worthwhile: he can afford to lose $50,000. It is clear that it is not sufficient to look at money only. Economists have therefore suggested that, instead of maximizing the expected profit, the decision-maker should maximize the expected *utility* of the outcome. Why this should be, and just what is meant by utility, is unfortunately beyond the scope of this book.

PROBLEMS ON EXPECTED VALUES

1. A fair die is tossed 100 (independent) times; let X be the number of aces obtained. Find the mean and variance of X.

2. A player tosses a fair coin 10 times. If the number X of heads is even, then the player wins X dollars. If X is odd, then the player loses X dollars. What are the player's expected winnings?

3. A merchant stocks units of a perishable item. Each unit costs him $2, and sells for $5. The merchant estimates that the demand k for this item has distribution $f(k)$ given by

$$f(0) = 0.1$$
$$f(1) = 0.15$$
$$f(2) = 0.25$$
$$f(3) = 0.3$$
$$f(4) = 0.15$$
$$f(5) = 0.05$$

How many units should the merchant stock to maximize his expected profits?

4. A fair coin is tossed 10,000 times. Using Chebyshev's inequality, find an upper bound for the probability that the number of heads obtained be under 4500 or over 5500.

5. If X and Y are random variables with mean μ_X and μ_Y respectively, and variance σ_X^2 and σ_Y^2 respectively, define the *covariance*

$$\sigma_{XY} = E[(X-\mu_X)(Y-\mu_Y)]$$

Prove that

$$\sigma_{XY} = E[XY] - E[X]E[Y]$$

and that, if $U = aX+b$, and $V = cY+d$, where a, b, c, and d are all constants, then

$$\sigma_{UV} = ac\sigma_{XY}$$

6. If X and Y are random variables with variance σ_X^2 and σ_Y^2 respectively, and covariance σ_{XY}, define the *correlation coefficient*

$$\rho_{XY} = \sigma_{XY}/\sigma_X\sigma_Y$$

Prove that, if $U = aX+b$, and $V = cY+d$, then

$$\rho_{UV} = \pm\rho_{XY}.$$

When will the negative sign hold here?

7. Let X, Y and Z be mutually independent random variables with variance σ_X^2, σ_Y^2 and σ_Z^2 respectively Let

$$U = X + Z$$
$$V = Y + Z.$$

Find σ_U^2, σ_V^2, σ_{UV} and ρ_{UV} in terms of σ_X^2, σ_Y^2 and σ_Z^2.

8. A fair coin is tossed 10 times. Let X be the number of heads, and let $Y = X^2$. Find the correlation coefficient ρ_{XY}.

9. An urn contains 100 balls each painted with two colors. One of the colors is either red or black; the other is either green or yellow. The distribution of colors is given by the following table.

	Green	Yellow
Red	38	15
Black	15	32

A ball is drawn from the urn. Let $X = 0$ if the first color is red, and 1 if it is black. Let $Y = 0$ if the second color is green, and 1 if it is yellow. Find the correlation coefficient ρ_{XY}.

10. A machine is either working well (probability 0.8) or not working well. If it is working well, its product will be good with probability 0.75; if it is not working well its product will be good with probability 0.4. Find the correlation coefficient ρ_{XY}, where $X = 0$ or 1 depending on whether the machine is working well or not, and $Y = 0$ or 1 depending on whether the product is good or defective.

12. MARKOV CHAINS

A Markov chain is a probabilistic process by which a system changes from one of several possible states to another at intervals of time. Examples of Markov processes include the way in which the length of a waiting line changes, the way in which a gambler's fortune grows, and the economic growth of a country. The characteristic property of such processes is that past history is unimportant, i.e. the future behavior of a system depends (up to a point) on the present state, but past sates will affect the future only in the sense that they have already affected the present. Thus, knowledge of the present state gives as much information (for matters of prediction) as knowledge of the system's entire history.

We shall assume that the system has only a finite number of possible states and that the transitions (changes from one state to another) can happen only at discrete intervals of time (say, every minute or every hour).

Let us suppose that a Markov process has n states. It is clear from our discussion that the process is entirely described by giving the n^2 probabilities of transition from one state to another (or to the same state); these n^2 numbers can be arranged to form an nth order matrix

$$A = (a_{ij}) \qquad i, j = 1, 2, \ldots, n$$

where a_{ij} is the conditional probability that the state will be in state j at time $t+1$, given that it is in state i at time t. The matrix A satisfies the conditions

3.12.1 $$a_{ij} \geq 0 \quad \text{for all } i, j$$

3.12.2 $$\sum_{j=1}^{n} a_{ij} = 1 \quad \text{for each } i$$

The reasons for conditions 3.12.1 and 3.12.2 should be clear. Any matrix that satisfies these conditions is known as a *stochastic matrix*.

III.12.1. Example. Two gamblers, G_1 and G_2, have a total of n units (of money) in their possession. They carry out a sequence of independent trials of a simple experiment with probability p of success. If the experiment is a success, G_2 pays one unit to G_1; if a failure, then G_1 pays one unit to G_2. This process terminates if one of the two gamblers is ever wiped out (i.e. has no units left).

This is a Markov process with $n+1$ states, the ith state ($i = 0, 1, ..., n$) coming when G_1 has i units. We see that, for $1 \leq i \leq n-1$, the system has a probability p of moving to state $i+1$, and q of moving to state $i-1$. For $i = 0$ or n, however, there are no further changes. The process is said to have an *absorbing barrier* at states 0 and n. We have, thus:

$$a_{i,i+1} = p \quad \text{for } i = 1, ..., n-1$$
$$a_{i,i-1} = q \quad \text{for } i = 1, ..., n-1$$
$$a_{00} = 1$$
$$a_{nn} = 1$$
$$a_{ij} = 0 \quad \text{for all other } i, j$$

For n = 5, the transition matrix would look like this:

$$\begin{bmatrix} 1 & 0 & 0 & 0 & 0 & 0 \\ q & 0 & p & 0 & 0 & 0 \\ 0 & q & 0 & p & 0 & 0 \\ 0 & 0 & q & 0 & p & 0 \\ 0 & 0 & 0 & q & 0 & p \\ 0 & 0 & 0 & 0 & 0 & 1 \end{bmatrix}$$

III.12.2. Example. Consider the following process: an urn contains n balls, k of which are red, and $n-k$, black. At each step, there is a probability rk that one of the red balls will become black, and, independently, a probability $r(n-k)$ (where r is a small number) that one of the black balls will become red.

Let the kth state hold when the urn has k red balls. There are n+1 possible states. To compute the transition probabilities, we note that we pass from state k to $k+1$ if one of the black balls "mutates" (i.e. changes color) but none of the red balls mutates. Therefore,

$$a_{k,k+1} = r(n-k)(1-rk)$$

Similarly, we have

$$a_{k,k-1} = rk(1 - r(n-k))$$

and, finally,

$$a_{k,k} = 1 - rn + 2r^2 k(n-k)$$

No other transitions are possible; i.e. it is not possible to pass in one step from state k to any state other than $k-1$, $k+1$, or k itself. Thus all other transition probabilities are zero.

For $n = 5$, and $r = 0.2$, the transition matrix would have the form

0	1	0	0	0	0
.04	.32	.64	0	0	0
0	.16	.48	.36	0	0
0	0	.36	.48	.16	0
0	0	0	.64	.32	.04
0	0	0	0	1	0

In dealing with Markov chains, it is often of interest to predict the state of the system, not in the time period that follows immediately, but some later period. It becomes natural to look for the "s-period" transition probabilities, $a_{ij}^{(s)}$: we define $a_{ij}^{(s)}$ as the probability that the system will be in state j at time $t+s$, given that it is in state i at time t. Then

$$A^{(s)} = (a_{ij}^{(s)})$$

is the *s-period transition matrix*.

It is clear that the matrices $A^{(s)}$ depend n the matrix A. Let us see how they are to be computed. What, for instance, is $a_{ij}^{(2)}$? We compute this by considering all the possible ways in which the system can reach state j at time $t+2$, given that it is in state i at time t.

Generally speaking, the system can pass from state i to some intermediate state k at time $t+1$, and then from k to j at time $t+2$. The probability of this is $a_{ik} a_{kj}$. Since k is arbitrary, we must add all these terms, obtaining

3.12.3
$$a_{ij}^{(2)} = \sum_{k=1}^{n} a_{ik} a_{kj}$$

But this is simply the rule for multiplication of matrices (in this case, for multiplying A with itself). We conclude that

$$A^{(2)} = AA = A^2$$

And, in general,

3.12.4 $\qquad A^{(s)} = A^s$

so that the s-period transition matrix is simply the sth power of the single-period transition matrix.

III.12.3. Example. The Markov chain with transition matrix

$$A = \begin{bmatrix} q & 1-q \\ 0 & 1 \end{bmatrix}$$

has the two-period matrix

$$A^2 = \begin{bmatrix} q^2 & 1-q^2 \\ 0 & 1 \end{bmatrix}$$

And, in general, it will have the s-period matrix

$$A^s = \begin{bmatrix} q^s & 1-q^s \\ 0 & 1 \end{bmatrix}$$

III.12.4. Example. The Markov chain with transition matrix

$$A = \begin{bmatrix} 0.3 & 0.6 & 0.1 \\ 0.1 & 0.5 & 0.4 \\ 0.4 & 0.1 & 0.5 \end{bmatrix}$$

has the 2-period matrix

$$A^2 = \begin{bmatrix} 0.19 & 0.49 & 0.32 \\ 0.24 & 0.35 & 0.41 \\ 0.33 & 0.34 & 0.33 \end{bmatrix}$$

and the 3-period matrix

$$A^3 = \begin{bmatrix} 0.234 & 0.391 & 0.375 \\ 0.271 & 0.360 & 0.369 \\ 0.265 & 0.401 & 0.334 \end{bmatrix}$$

13. REGULAR AND ABSORBING MARKOV CHAINS

For most Markov chains, the state in the near future will depend considerably on the present state. On the other hand, it is not inconceivable that (for many chains at least) the importance of the present state should vanish over a long period of time. For example, if we see a long line at a supermarket checkout counter, we should be surprised to find that, only a few minutes later, the line had vanished. On the other hand, a long line in the morning should probably have little influence on thew length of the line in the evening. This type of behavior seems evident in the chain of Example III.12.2 above, where the system's characteristics tend to push it away from the extremes of too many red or too many black balls. Example III.12.4 also seems to behave in this manner: note how the three rows, which were very dissimilar in A, are closer to each other in A^3. The pattern is even more obvious in the 4-period matrix

$$A^4 = \begin{array}{|c|c|c|} \hline 0.259 & 0.374 & 0.367 \\ \hline 0.265 & 0.380 & 0.355 \\ \hline 0.253 & 0.393 & 0.354 \\ \hline \end{array}$$

Further multiplication would show that, for large s, the rows of A^s are almost equal; in fact, it would be seen that they approach the matrix

$$V = \begin{array}{|ccc|} \hline 0.259 & 0.382 & 0.359 \\ 0.259 & 0.382 & 0.359 \\ 0.259 & 0.382 & 0.359 \\ \hline \end{array}$$

Thus, for large s, the state at time $t+s$ is almost independent of the state at time t. The vector $\mathbf{v} = (0.259, 0.382, 0.359)$ represents the "steady-state" behavior of the system; i.e. in the distant future, the probabilities that the system be in any one of the three states are given by this vector. Note that it is a solution of the equation $\mathbf{v}A = \mathbf{v}$.

In fact, this convergence of the matrices A^s (to a matrix with all rows equal) occurs quite frequently. It will certainly happen if all the entries in the matrix A are positive (as opposed to zero).

In fact, suppose all the $a_{ij} > 0$. Let a be the smallest of these:

$$a = \min_{i,j} a_{ij}$$

and let $r = 1-a < 1$. Now, let $d(s)$ be the largest difference between entries in the same column of the matrix A^s:

$$d(s) = \max\{a_{ij}^{(s)} - a_{kj}^{(s)}\}$$

Intuitively, it is easy to see that, if $d(s)$ is small, then the rows of A^s are all nearly equal.

We use now the fact that $A^{s+1} = AA^s$ and so

$$a_{ij}^{(s+1)} = \sum a_{il} a_{lj}^{(s)}$$

We know that A is a stochastic matrix and so

$$\sum_{l=1}^{n} a_{il} = 1$$

Let $a_{pj}^{(s)}$ be the largest, and $a_{qj}^{(s)}$ the smallest, of the entries in the jth column of A^s. We have then

$$a_{ij}^{(s+1)} = a_{ip} a_{pj}^{(s)} + \sum_{l \neq p} a_{il} a_{lj}^{(s)}$$

Now, $a_{ip} \geq a$ and, for all l, $a_{lj}^{(s)} \geq a_{qj}^{(s)}$. It will follow that, for any i,

3.13.1 $\qquad a_{ij}^{(s+1)} \geq a\, a_{pj}^{(s)} + (1-a)\, a_{qj}^{(s)}$

In a similar manner, we can see that, for any k,

3.13.2 $\qquad a_{kj}^{(s+1)} \leq (1-a)\, a_{pj}^{(s)} + a\, a_{qj}^{(s)}$

and so

$$a_{kj}^{(s+1)} - a_{ij}^{(s+1)} \leq r\,(a_{pj}^{(s)} - a_{qj}^{(s)}).$$

Since this holds for all i, j, and k,

3.13.3 $\qquad d(s+1) = \max\{a_{ij}^{(s+1)} - a_{kj}^{(s+1)}\} \leq rd(s).$

We conclude several things from this. First, from (3.13.1) and (3.13.2), each entry in the jth column of A^{s+1} is at least as large as the smallest entry in the jth column of A^s and no larger than the largest entry in the jth column of A^s. From (3.13.3), we conclude, $d(s) \leq r^s$, and so, as s increases, $d(s)$ goes to zero, so that the rows of A^s are nearly equal to each other and also to the rows of A^{s+1} and all higher powers of

A. It will follow from all this that the matrices A^s approach a limiting matrix, V, all of whose rows are equal. The rows of V represent the "steady state" behavior of the Markov chain.

In general, it is not necessary that all entries in A be positive for this type of convergence to hold. It suffices that for some power of A, all entries be positive.

III.13.1. Definition. A stochastic matrix, A, is *regular* if at least one of its powers, A^s, has all positive entries.

We give the following theorem without further proof:

III.13.2. Theorem. Let A be a regular matrix. Then
(a) The matrices A^s approach a matrix V, as s grows.
(b) All rows of V are equal to a vector **v**.
(c) $vA = v$.

Generally speaking, the easiest way to find the limit matrix V is to solve the equation $vA = v$. If A is regular, this equation will have only one solution which also satisfies the stochastic condition

$$\sum v_j = 1$$

III.13.3. Example. The cigarette company that manufactures brand C starts an aggressive (and very successful) advertising campaign. The results of this campaign are such that, of people smoking brand C in a given week, 80 per cent continue to smoke it in the following week (and 20 per cent switch to other brands). Of those smoking any other brands in a given week, 40 per cent are won over to brand C the following week. In the long run, what fraction of smokers will be smoking brand C?

We have here a Markov chain, with transition matrix

$$A = \begin{array}{|cc|} 0.8 & 0.2 \\ 0.4 & 0.6 \end{array}$$

We consider here the equation $\mathbf{v}A = \mathbf{v}$, or

$$0.8v_1 + 0.4v_2 = v_1$$
$$0.2v_1 + 0.6v_2 = v_2$$

This has the solution

$$v_1 = 2v_2$$

Since we want $v_1 + v_2 = 1$, this gives us

$$\mathbf{v} = (2/3, 1/3)$$

It follows that

$$V = \begin{bmatrix} 2/3 & 1/3 \\ 2/3 & 1/3 \end{bmatrix}$$

In the long run the company will have 2/3 of the market. Note that the initial state does not matter: even starting from scratch, the company will, eventually, "nearly" capture its share of the market (though, of course, starting from scratch, it will take longer to do so).

If the transition matrix is not regular, there is still a possibility that the process may behave in this manner (see Example III.12.3), but in general this need not happen. A special case arises when some of the states in the system form *absorbing barriers*: once a system reaches such a state, there is no further change. Example III.12.1 has this property, inasmuch as the process terminates as soon as one of the two gamblers is wiped out, so that states 0 and n are absorbent states. It is not too difficult to see (in this example) that, if s is large enough, then the game will "very probably" have terminated before s trials of the experiment: one of the two gamblers will have been wiped out. It may be shown (by an argument similar to the proof of Theorem III.13.2) that, as s increases, the matrices A^s approach a limit matrix, W, which has non-zero entries only in the first and last columns ($j = 0$ or n). The entry w_{i0} is then the probability that gambler G_1 will eventually be wiped out, given that he starts with i units. Unfortunately, we cannot study this problem here.

PROBLEMS ON MARKOV CHAINS

1. When a company president finds an item on his desk, he may:
   (a) send it to the vice-president, to act on it the following day;
   (b) leave it on his own desk for the following day, or
   (c) act on it immediately.

   The probabilities of these three options are 0.3, 0.6, and 0.1 respectively.

   Similarly, the vice-president has the options of keeping the paper on his own desk, sending it to the president, and acting on it immediately; these options have probability 0.5, 0.2, and 0.3 respectively.

   Represent this as a Markov chain, and form a transition matrix. If the president has the paper, what is the probability that some action will be taken within three days?

2. Three soap brands, I, II, and III, dominate their field. There is a continuous switching by customers, which can be represented probabilistically. If a customer is now using brand I, there is a 0.6 probability that he will continue to use it the following week, 0.3 probability that he will switch to brand II, and 0.1 that he will switch to III. If he is using brand II, there is 0.5 probability that he will continue to use II, 0.4 that he will switch to I, and 0.1 that he will switch to III. If he is using brand III, there is 0.7 probability that he will continue to use it, and 0.3 that he will switch to II.

   Given that a consumer is now using brand II, what is the probability that He will be using brand I, three weeks from now?

3. Using the data of problem 2, assume that the brands I, II, and III are presently holding 50 per cent, 30 per cent, and 20 per cent, respectively, of the market. What shares will they hold two weeks from now?

4. Again using the data of problem 2, what share of the market will the three brands hold in the long run?

5. Customers arrive at random at a service counter that can hold four people. In an interval of time, there is a probability 0.3 that a new customer will arrive (though he will be turned away if there are already four present) and a 0.2 probability that the customer at the head of the line (if there is such a customer) will finish service and leave. The probabilities that more than one person will arrive or that more than one will finish service (in such an interval) are considered negligible.

(a) Express this situation as a Markov chain, in which the state is the number of persons at the counter.
(b) What is the probability that the counter be idle (i.e. no customers present)?
(c) What is the probability that a customer be turned away?

5. Consider a Markov chain with $m+n$ states, $S_1, S_2, \ldots, S_{m+n}$. Suppose that, for $n+1 \leq i \leq m+n$, $a_{ii} = 1$; i.e. if the chain reaches one of the last m states, it will remain thereafter in that state. Suppose, moreover, that for $i = 1, \ldots, n$,

$$t_i = \sum_{j=n+1}^{m+n} a_{ij} > 0$$

i.e., if the chain is in one of the first n states, then there is a positive probability that it will pass to one of the last m states. Then the last m states are *absorbing* states, while the first n states are *transient* states. Show that, as k approaches infinity, the powers A^k of the transition matrix approach a matrix, W, which is such that $w_{ij} = 0$ for $1 \leq j \leq n$. (Essentially, this means that, with probability 1, the Markov process will eventually reach one of the absorbing states and remain there.) Prove this by showing that, if t is the minimum of the t_i, then the entries in the first n columns of A^k are all smaller than or equal to $(1-t)^k$. Since $t > 0$, these powers all approach 0.

7. Show that the conclusions of Problem 6 will hold if it is possible to reach one of the absorbing states from any one of the transient states in not more than p moves, where p can be any integer. (Do this by considering the p-period transition matrix A^p.)

8. With the assumptions of problem 6, let $B = (b_{ij})$ be the $n \times n$ matrix formed by the first n rows and columns of A. Then B is called the *transient part* of A. The matrix

$$C = (I-B)^{-1}$$

(which may be shown to exist) is known as the *fundamental matrix* of the chain. Show that, if k is large, then C is approximately equal to

$$I + B + B^2 + B^3 + \ldots + B^k$$

And that the entries c_{ij} represent the expected number of periods which the chain will spend in the jth transient state (before absorption) given that it is currently in the ith state.

9. With the assumptions of Problem 6, let d_i be the expected number of periods before the system reaches an absorbing state, given that it is now in state i. Show then that, for $n+1 \leq i \leq m+n$,

$$d_i = 0$$

whereas, for $1 \leq i \leq n$,

$$d_i = 1 + \sum p_{ij} d_j$$

10. Two gamblers, with a total of 5 units between them, play a fair game (i.e. one with a probability 0.5 of winning) for stakes of 1 unit each time. The game will terminate whenever either of the two gamblers holds all 5 units.
   (a) Given that one gambler holds 2 units now, what is the expected number of times that he will hold 3 units, before the game terminates?
   (b) What is the expected duration of the game?
   (c) What is the probability that he will eventually be wiped out?

IV. THE THEORY OF GAMES

1. GAMES: EXTENSIVE AND NORMAL FORM

The theory of games is a branch of mathematics designed to analyze the behavior of people in situations of conflict of interest. The exact nature of these conflicts is best evaluated in the case of parlor games (in which such things as winning and losing are clearly defined) and it is this fact that has given the theory its name. Notwithstanding, the theory is applicable to many serious situations: a *game*, as the term shall be used here, may be an extremely serious affair, including such things as cutthroat economic competition and nuclear warfare.

Research into the theory of games was started independently in the 1920's by J. von Neumann and E. Borel. Some small progress was made in the 1930's, but it was not until the Second World War that it reached the status of an independent branch of mathematics (closely allied to economics); von Neumann and O. Morgenstern collaborated on its rigorous formulation. Since then a close relationship between game theory and linear programming has contributed to its development; progress has been rapid.

The basic elements of a game can be seen in parlor games such as bridge, poker, or chess. These include:

(a) An alternation of moves, some of which may be random moves (such as shuffling a pack of cards), while others are personal moves.
(b) Well-defined probability distributions for the random moves (note that attempts to change these distributions, as by stacking a deck or loading dice, are considered "cheating".
(c) Some knowledge (which may be perfect, as in chess, or very poor, as in poker) of the outcomes of previous moves.
(d) A function that assigns payoffs (in terms such as money, fame, or satisfaction) to each terminal position, or *play*, of the game.

A strict mathematical definition of this structure can be given, but would serve no purpose here. For this, the interested reader is referred to the literature.

It is generally possible to represent a game by a graph (a set of arcs and nodes); this is usually known as the *game tree*, or *extensive form* of the game. The nodes (vertices) of the graph represent the many possible positions of a game; the arcs represent the moves. There will be an arc from node A to node B if the game can be changed from position A to position B in a single move. An elementary, but instructive, example is the game of "matching pennies."

IV.1.1. Example (Matching Pennies). A player, P_1, puts a coin either "heads up" (H) or "tails up" (T). A second player, P_2, in ignorance of P_1's choice, calls either "heads" or "tails". If P_2 guesses correctly (i.e. if their choices match), then P_1 pays 1 unit to P_2; otherwise P_2 pays 1 unit to P_1.

FIGURE IV.1.1. Game tree for "matching pennies."

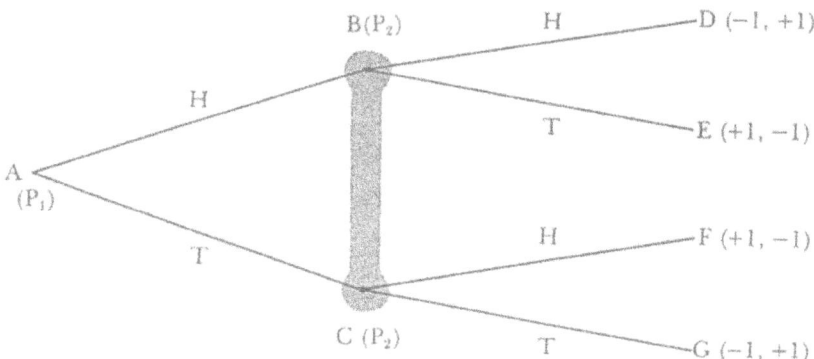

Figure IV.1.1 shows the game tree for matching pennies. Each of the vertices is labeled to designate the player who moves at the given position. Vertex A "belongs" to P_1, while B and C "belong" to P_2. The other vertices represent terminal positions and thus don't belong to either player. Instead, a payoff vector is assigned to each of them. Vertex A is "distinguished": it represents the initial position. Note the connection between vertices B and C: this represents the fact that P_2 will not know whether he is at B or at C; the two vertices are then said to belong to the same *information set*.

A second example is a (very rudimentary) form of poker:

IV.1.2. Example. A deck contains $m+n$ cards, m of which are marked "high" (H) and n "low" (L). One of these cards (chosen at random) is given to player P_1; he can then "fold", in which case he pays 1 unit to P_2, or bet a units. If P_1 bets, then P_2 can fold, in which case he pays P_1 1 unit, or "call" (i.e. match the bet). After a call, P_1 pays P_2 $1+a$ units if the card is L, and P_2 pays P_1 $1+a$ units if the card is H.

Figure IV.1.2 shows the tree for this game. Vertex A is a chance move.

Note that a probability distribution ($m/(m+n)$ for H and $n/(m+n)$ for L) is given at this vertex. Note also how vertices E and F are connected: they belong to the same information set, since P_2 does not know whether P_1 really has a high card (vertex E) or is bluffing (vertex F).

FIGURE IV.1.2. A rudimentary form of poker (Example IV.1.2).

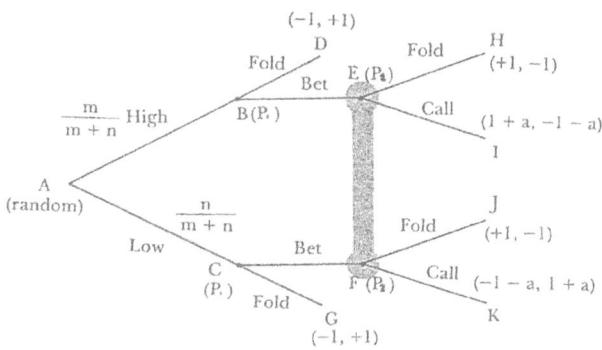

In our attempt to analyze these games, we shall assume, of course, that each player tries to maximize his payoff, given in units of money or possibly of utility. This is complicated by the fact that certain games involve randomizations, so that, even if we know what the players choose to do, we still do not know, except in probability, what the outcome of the game will be. We shall overcome this difficulty by assuming (somewhat unjustifiably, perhaps) that the players will attempt to maximize their *expected* returns. It will then suffice to deal with expected values.

Let us suppose that some persons wish to play a game but, unfortunately, can find no time that is convenient for all of them. If the game is not too difficult, it is possible for them to play it by proxy: each one can give detailed instructions to his secretary. Then, if the instructions are comprehensive enough, it will be possible for each secretary to play the game without needing to make a single decision – since the decisions all appear in the instructions given by the original players. Such a comprehensive set of instructions – one that tells a player what to do in each conceivable position of a game – is known as a *strategy*. For instance, we see that in Example IV.1.1 each player has two strategies, which we label H and T, respectively. In Example IV.1.2, P_1 has four strategies, since he can distinguish two positions (vertices B and C) and has two alternatives at each of them. These four strategies are: (a) to bet in any case; (b) to fold in any case; (c) to bet with H, and fold with L; and (d) to fold with H, and bet with L. On the other hand, P_2 has only two strategies: he can call a bet or fold. Note that he would like to be able to call bluffs only, but that is not possible since he cannot distinguish between an honest bet and a bluff. (If he could, he would have four strategies.)

We will not generally deal with the game trees (extensive form of the game). Instead, we prefer a more concise form.

IV.1.3. Definition. The *normal form* of a game is a list of all strategies for all the players, together with the expected payoffs for each possible choice of strategies (one for each player).

In the two examples just given, the amounts won by one player are lost by the other player; i.e. the games represent a closed system. Such games are called *zero-sum games*, since the sum of the amounts won by the two players (treating a loss as a negative win) is always zero. A zero-sum game for two players is also known as a *strictly competitive game*, since the situation is such as to preclude cooperation by the two players. If, in a strictly competitive game, each of the two players has only a finite number of strategies, then the normal form of the game can be given as a matrix $A = (a_{ij})$: each row represents a strategy for P_1 and each column represents a strategy for P_2. The entry a_{ij} will then be the expected amount won by P_1 (and thus lost by P_2), assuming that they choose their ith and jth strategies respectively. Because of this, such games are known as *matrix games*. As examples, we give the game matrices for the two games just described.

IV.1.4. Example. For the game described in Example IV.1.1, each of the two players has two strategies, H and T. The game matrix is

	H	T
H	-1	1
T	1	-1

(We repeat that positive entries mean gains for P_1, negative entries are losses for P_1.)

IV.1.5. Example. For the rudimentary poker game of Example IV.1.2, let us first simplify notation by letting $p = m/(m+n)$ be the probability of a high card, and $q = n/(m+n)$ be the probability of a low card.

The first player has four strategies which we call BB (bet in any case), BF (bet on H, fold on L), FB (fold on H, bet on L), and FF (fold in any case). The second player has the strategies C (call) or F (fold). Now, if P_1 chooses BB and P_2 chooses

C, then the payoff (to P_1) will be $1+a$ if he has H (probability p) and $-1-a$ if he has L (probability q). The expected payoff then is

$$p(1+a) + q(-1-a) = (p-q)(1+a).$$

If P_1 chooses BB and P_2 chooses F, then P_1 will always win 1 unit. In a similar way, we can compute the expected payoffs in all other case, to obtain the matrix

	C	F
BB	$(p-q)(1+a)$	1
BF	$p(1+a) - q$	$p-q$
FB	$-1- qa$	$q-p$
FF	-1	-1

2. SADDLE POINTS

Let us consider the following matrix game (in which, as before, the entries represent payoffs to the column player, P_2, to the row player, P_1):

2	2	1	4
0	2	5	3
4	2	3	2

We shall try to analyze the game by (attempting to) duplicate the two players' thinking process. We assume, as before, that P_1 wishes to maximize the payoff, while P_2 wishes to minimize it.

Let us consider first P_2's point of view. He could, of course, choose the second column, which guarantees that he loses exactly 2 units. On the other hand, by choosing the first or third columns, he might hope to lose nothing at all, or only one unit. What should he do?

At first, this is not clear. It does seem quite clear that he should not choose the fourth column, since the second column is distinctly better. In fact, we see that, whatever P_1 chooses, P_2 will always do at least as well, and sometimes better, if he chooses the second column rather than the fourth. (This is because each entry in the second column is smaller than or equal to the corresponding entry in the fourth column.) P_2 will therefore discard the fourth column and consider only the other three. This gives us a reduced game

$$\begin{bmatrix} 2 & 2 & 1 & \cancel{4} \\ 0 & 2 & 5 & \cancel{3} \\ 4 & 2 & 3 & \cancel{2} \end{bmatrix}$$

Let us now consider P_1's point of view. He knows that, assuming his opponent to be rational, the fourth column can be discarded. Looking only at the remaining columns, he can see that the first row can be discarded in favor of the third. In fact, each entry in the third row is larger than or equal to the corresponding entry in the first row. This leaves us a smaller matrix

$$\begin{bmatrix} \cancel{2} & \cancel{2} & \cancel{1} & \cancel{4} \\ 0 & 2 & 5 & \cancel{3} \\ 4 & 2 & 3 & \cancel{2} \end{bmatrix}$$

In this 2×3 matrix, P_2 sees that he can discard the third column, which is uniformly worse than the second. This leaves

$$\begin{bmatrix} \cancel{2} & \cancel{2} & \cancel{1} & \cancel{4} \\ 0 & 2 & \cancel{5} & \cancel{3} \\ 4 & 2 & \cancel{3} & \cancel{2} \end{bmatrix}$$

Now it is P_1's turn: in this matrix, the last row is at least as good as the other, and some times better. We will be left with

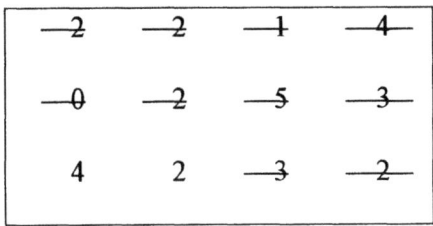

Thus, if P_1 is rational, and if he assumes that his opponent is rational, then he will always choose the bottom row of the matrix. This means that P_2 should always choose the second column.

We see, then, that a very rational analysis of the situation will always lead P_1 to choose the third row, while P_2 will choose the second column. This is essentially a very pessimistic point of view: each player assumes that the other is capable of carrying out this analysis and acting rationally on this basis. If P_1 were very optimistic, he might decide to choose the second row (which gives him a chance to win 5 units if P_2 slips up); similarly, an optimistic P_2 might choose the first column, hoping that P_1 might make a poor choice. Game theory is pessimistic, at least to this extent: opponents are assumed to be rational and intelligent.

The method by which we have deleted rows and columns is known as *row* and *column domination*.

IV.2.1. Definition. Let $A = (a_{ij})$ be a matrix game. We say that the *i*th row *dominates* the *k*th row if $a_{ij} \geq a_{kj}$ for each j and $a_{ij} > a_{kj}$ for at least one value of j. Similarly, the *j*th column dominates the *l*th column if $a_{ij} \leq a_{il}$ for all i and $a_{ij} < a_{il}$ for at least one value of i.

In essence, one strategy dominates another strategy (of the same player) if it is never worse and is sometimes better. The importance of this notion is that dominated strategies can always be discarded: optimal behavior will never require them.

IV.2.2. Example. In the game with matrix

3	1	5
2	0	6

the second column dominates both of the other columns. Once the first and third columns are deleted, the first row dominates the second row (note that this does not happen if the third column is not deleted). We conclude that P_1 should choose the first row, while P_2 chooses the second column.

Helpful as this method of domination may be, it is not always available for us. Nevertheless, a similar type of reasoning will help us in certain other cases. Consider for example the matrix game

6	2	1
4	3	5
0	1	6

As before, we look for row and column dominations, but there are none. We cannot reduce the matrix in this fashion. Let us, nevertheless, analyze the game from a pessimistic point of vie

Let us suppose P_1 chooses his first strategy. Then, depending on what P_2 does, P_1 will win 6, 2, or 1 unit. A pessimistic P_1 will consider the worst that can happen to him: he can count on winning 1 unit. This is the *security level* that his first strategy will give him. In a similar way, his second strategy gives him a security level of 3 units, while the third strategy gives him a security level of 0 units. We see that P_1 can maximize his security level by choosing the second row (strategy).

Analyze the game from P_2's point of view. If he chooses his first strategy, he might lose 6, 4, or 0 units. His security level (the worst that can happen to him) is 6 units lost. Similarly, his second and third columns (strategies) yield security levels of 3 and 6 units lost, respectively. We conclude that his best choice, as far as security level is concerned, is the second column.

Let us see what this means. P_1 has a "gain-floor" of 3 units, obtained by using the second row. P_2 has a "loss-ceiling" of 3 units, obtained with the second column. In other words, P_1 can assure himself a gain of 3 units, while P_2 can insure himself against any greater loss. This being so, it seems reasonable to demand that, for this game, rational play by intelligent players should bring about this payoff of 3 units. P_1 will choose the second row, while P_2 chooses the second column. Nor is this all: if P_1 suspects that P_2 will choose the second column, this will give him more reason to choose the second row. Conversely, if P_2 suspects that P_1 will choose the second row, this in turn will give him more reason to choose the second column. In other words, the arguments in favor of this outcome reinforce each other. The outcome is, in this, sense, determined: we will say that the game *is strictly determined* and has a *value* of 3 units.

The fundamental property in this game is that the two strategies (second row, second column) are in *equilibrium*: neither player can gain by a unilateral change from these strategies. This is clear since the entry corresponding to these two strategies is both the largest in its column (so that P_1 cannot gain by changing rows) and the smallest in its row (so that P_2 cannot gain by changing columns). Such an entry is known as a *saddle point*.

IV.2.2. Definition. Let A be a matrix game. The entry a_{ij} is a saddle point if, for any k,

4.2.1 $$a_{ij} \geq a_{kj}$$

and, for any l,

4.2.2 $$a_{ij} \leq a_{il}$$

The name, saddle point, comes by analogy from the surface of a (horse) saddle, which curves up in one direction and down in another (see Figure IV.2.1).

FIGURE IV.2.1. A saddle-shaped surface.

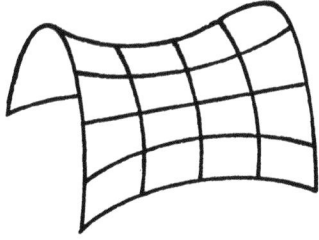

Our analysis suggests that if a game has a saddle point, both players should choose the strategies which lead to this point. A question naturally arises about games with two (or more) saddle points. It is not inconceivable that, for such a game, P_1 might try for one saddle point, while P_2 tries for another, and (conceivably) something very different from a saddle point might arise. While this type of pathology does sometimes arise in game theory, it cannot happen with matrix games: this is shown in the following theorem.

IV.2.3. Theorem. Let a_{ij} and a_{kl} be saddle points for the matrix game A. Then a_{kj} and a_{il} are also saddle points and, moreover,

$$a_{ij} = a_{kl}.$$

Proof. Since a_{ij} is a saddle point, we must have

$$a_{ij} \geq a_{kj}$$
$$a_{ij} \leq a_{il}$$

Since a_{kl} is also a saddle point,

$$a_{kl} \geq a_{il}$$
$$a_{kl} \leq a_{kj}$$

Thus we have

$$a_{ij} \geq a_{kj} \geq a_{kl} \geq a_{il} \geq a_{ij}$$

But this means

$$a_{ij} = a_{kj} = a_{kl} = a_{il}$$

Now, a_{ij} is the smallest entry in the ith row, while a_{kl} is the largest entry in the lth column. The entry a_{il} is in the ith row and in the lth column, and is equal to both a_{ij} and $= a_{kl}$. But this means a_{il} is the smallest entry in its row and column, and is therefore a saddle point. Similarly, a_{kj} is also a saddle point.

Theorem IV.2.3 shows that saddle points for matrix games are always *equivalent* (they give the same payoff) and *interchangeable,* in the sense that if one player aims for one saddle point, and the other aims for another saddle point, the result is nonetheless a saddle point. Thus the possible multiplicity of saddle points is not a problem: if both players want saddle points, the outcome will always be the same.

PROBLEMS ON SADDLE POINTS

1. Solve the following matrix games:

(a)
3	5
4	6

(b)
3	6	4
1	0	8

(c)
2	0	6
5	1	3
7	-2	1

(d)
3	5	1	2
4	8	3	0
5	4	3	6

(e)
7	6	1	2
5	5	4	6
4	8	3	0

2. Two competing companies must build their next branch at one of three cities, A, B, and C, whose location and distances are shown in Figure IV.2.2. If both companies build in the same city, they will split the business evenly. If they build in

different cities, then the company that is closer to a given city will get all of that city's business. If all three cities have the same amount of business, where should the companies build?

FIGURE IV.2.2.

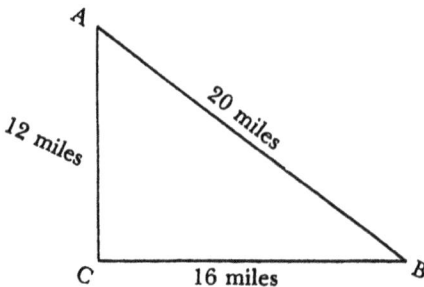

3. Show that, if all the entries in a 2×2 matrix are positive, and the matrix is not invertible, then the matrix has a saddle point.

4. Show that, in a 2×2 matrix, if both entries in one row are equal, then the matrix has a saddle point.

5. Show that, if a 2×2 or 2×3 matrix has a saddle point, then either one row dominates the other, or one column dominates another (or possibly both). Show by a counterexample that this is not true for 3×3 matrices.

3. MIXED STRATEGIES

In the previous section we studied the properties of saddle points; arguments were given for the use of saddle point, or equilibrium, strategies by both players. Unfortunately, only comparatively few games have saddle points. For example, a game as simple as matching pennies (Examples IV.1.1 and IV.1.4), with matrix

$$\begin{array}{|cc|} \hline -1 & 1 \\ 1 & -1 \\ \hline \end{array}$$

Does not have a saddle point. For such a game, it is difficult to see what either player should do. P_1 can obtain a gain-floor of -1 unit, while P_2 can obtain a loss-ceiling of 1 unit, but these both represent definite losses. Nor does it help to try to outguess the opponent: P_1 can be hurt just as much by crediting P_2 with too much intelligence as by not crediting him enough. In fact, we can imagine P_1 reasoning along these lines: "People generally choose H, so P_2 will expect me to choose H, and so he himself will choose H. Therefore I should choose T. But perhaps he has reasoned that way and expects me to choose T. Therefore I should choose H. But perhaps that is P_2's reasoning, too, and so I should choose T. But then...." It is clear that this sort of thing will get the players nowhere. The difficulty, in brief, is this: in order to be "good" (whatever meaning may be attached to this word), it must be rationally chosen. But if a player chooses rationally, his opponent (who is also rational) can reproduce his thinking process and guess what he plans to do. Yet the essence of a game such as this is, precisely, a need for secrecy: a player loses if his opponent discovers what he plans to do. This is in sharp contrast to the situation for games with saddle points, in which knowledge of the other's plans does not cause the players to change their own. (The chief trouble is that there is no equilibrium here.)

The apparent conflict between rationality and unpredictability was solved, almost simultaneously, by von Neumann and Borel. In essence their suggestion was quite simple: it had been in use, more or less unconsciously, by most gambling establishments. The suggestion was to choose the strategy by means of some random mechanism. There was, however, a new element introduced: the random mechanism would behave according to a rationally chosen scheme. In other words, the probabilities of each of the strategies should be rationally chosen. But, while our opponent might be able to figure out the probabilities in the randomization scheme (by somehow reproducing our manner of thinking) he cannot decide which strategy is to be used at each p0lay of the game (since the randomization mechanism acts irrationally). For a poker player, the question might be: granted that one needs to bluff occasionally, and to do so at random, how frequently (i.e. with what probability) should one bluff?

This gives us the idea of a *mixed strategy*: it is a probability distribution on the set of a player's strategies (henceforth called *pure strategies* to distinguish them from mixed strategies).

IV.3.1. Definition. Let A be an $m \times n$ matrix game. Then, a mixed strategy for P_1 is a vector $\mathbf{x} = (x_1, x_2, ..., x_m)$ satisfying

4.3.1
$$\sum_{i=1}^{m} x_i = 1$$

4.3.2 $$x_i \geq 0 \quad i = 1, 2, \ldots, m$$

Similarly, a mixed strategy for P_2 is a vector $\mathbf{y} = (y_1, y_2, \ldots, y_n)$ satisfying

4.3.3 $$\sum_{j=1}^{n} y_j = 1$$

4.3.4 $$y_j \geq 0 \quad j = 1, 2, \ldots, n$$

For the game of "matching pennies," described earlier, the obvious suggestion is to decide on H or T by tossing a coin. Assuming the coin to be fair, this gives the mixed strategy (1/2, 1/2) for P_1. It is easy to see that, whatever P_2 does, the game will have an expected payoff 0. In other words, use of a mixed strategy gives P_1 a gain-floor of 0 as compared to the gain-floor of -1 that he can obtain by using pure strategies. In a similar way, P_2 can use the mixed strategy (1/2, 1/2), which gives him a loss-ceiling of 0 compared to the loss-ceiling of 1 obtained by using pure strategies. We see that the gain-floor and loss-ceiling are equal; it follows that the two mixed strategies which guarantee the value 0 are in equilibrium and we say that the game has a *saddle point in mixed strategies*. Thus, if we admit mixed strategies, and agree to use the expected utility as objective (as we earlier said that we would), it follows that this game is almost as well determined as the games studied in Section 2 of this chapter. Both players, then, should use the mixed strategies (1/2, 1/2); moreover, knowledge of the opponent's mixed strategy should not cause either player to change his own.

The question naturally arises as to whether every game will have a saddle point in mixed strategies. The answer is affirmative. We shall leave its proof until later; first, we give an analysis of the situation.

Let us assume that, for the matrix game A, player P_1 chooses the mixed strategy $\mathbf{x}$, while P_2 chooses $\mathbf{y}$. Since the two randomizations are independent, the probability that P_1 chooses his ith (pure) strategy while P_2 chooses his jth (pure) strategy will be $x_i y_j$. The payoff in this case is a_{ij}, and so the expected payoff will be

4.3.5 $$A(\mathbf{x}, \mathbf{y}) = \sum_{i=1}^{m} \sum_{j=1}^{n} a_{ij} x_i y_j$$

or, in matrix notation,

4.3.6 $$A(x, y) = xAy'.$$

Let us suppose, now, that P_1 is considering using the mixed strategy **x**. As before, we assume that he is cautious; therefore he will at least look at the worst that can happen to him with such a strategy. His security level, then, is

4.3.7 $$u(x) = \min_{y \in Y} xAy'$$

where the minimum is taken over all possible mixed strategies, **y**, for P_2.
  Similarly, if P_2 uses the mixed strategy **y**, his security level is

4.3.8 $$w(y) = \max_{x \in X} xAy'$$

where the maximum is taken over all possible mixed strategies, **x**, for P_1.

It is not difficult to see that, if **x** and **y** are mixed strategies for P_1 and P_2, respectively, then

4.3.9 $$u(x) \leq w(y).$$

In fact, P_1 can assure himself of $u(x)$, while P_2 can insure himself against losing more than $w(y)$. Thus (4.3.9) must hold. Suppose, however, that there exist strategies **x*** and **y*** such that

4.3.10 $$u(x^*) = w(y^*) = v$$

In that case, it is clear that **x*** and **y*** are *optimal* in the sense that they maximize and minimize, respectively, the two players' security levels. (It may be noted that the situation is similar here to that we found when we were dealing with dual linear programs; that this is indeed the case will be established.) The number v, which is the common value of the two players' security levels, will be known as the *value* of the game, and it is clear that, in expectation, both players can guarantee the value of the game by using the mixed strategies **x*** and **y*** respectively. Neither can hope to gain by switching unilaterally, the two strategies being in equilibrium.

  Let us consider the expression (4.3.7). This defines $u(x)$ as the minimum of the terms xAy'. This minimum can best be found if we write

$$\sum_{i=1}^{m}\sum_{j=1}^{n} a_{ij} x_i y_j = \sum_{j=1}^{n} y_j \{\sum_{i=1}^{m} a_{ij} x_i\}$$

It may be seen that $\sum a_{ij} x_i$ is the jth component of the vector $\mathbf{x}A$. We shall write this as $(\mathbf{x}A)_j$, and so

$$\mathbf{x}A\mathbf{y}^t = \sum_{j=1}^{n} y_j (\mathbf{x}A)_j$$

so that $\mathbf{x}A\mathbf{y}^t$ is an average (i.e., expected value) of the components of the vector $\mathbf{x}A$. It will follow that (for a given x), this expectation $\mathbf{x}A\mathbf{y}^t$ can be made as small as the smallest component of $\mathbf{x}A$, and no smaller. In other words, the security level $u(\mathbf{x})$ is the smallest component of the vector $\mathbf{x}A$. (This suggests that P_2 can get along with using pure strategies only, as, in fact, he could if he knew P_1's choice of strategy. But P_2 must also use mixed strategies to prevent P_1 from finding out.) Thus,

4.3.11 $$u(\mathbf{x}) = \min_{j} (\mathbf{x}A)_j$$

and, similarly,

4.3.12 $$w(\mathbf{y}) = \max_{i} (A\mathbf{y}^t)_i$$

where $(A\mathbf{y}^t)_i$ represents the ith component of the vector $A\mathbf{y}^t$. This leads us to the following definitions:

IV.3.2. Definition. Let A be a matrix game. We say that v is the *value* of A if there exist strategies $\mathbf{x}^*$ and $\mathbf{y}^*$ (for P_1 and P_2 respectively) such that

4.3.13 $$\sum_{i=1}^{m} a_{ij} x_i \geq v \quad \text{for } j = 1, 2, \ldots, n$$

and

4.3.14 $$\sum_{j=1}^{n} a_{ij} y_j \leq v \text{ for } i = 1, 2, \ldots, m$$

In this case the strategies **x*** and **y*** are *optimal strategies*. A *solution* of the game is a triple $(v, \mathbf{x^*}, \mathbf{y^*})$ consisting of the value and a pair of optimal strategies (one for each player).

In terms of matrix operations, the conditions (4.3.13) and (4.3.14) can be restated in the form

4.3.15 $$\mathbf{x}A \geq v\mathbf{J}$$

4.3.16 $$A\mathbf{y}^t \leq v\mathbf{J}^t$$

where J is the row vector $(1, 1, \ldots, 1)$ of desired size (n components in (4.3.15) and m components in (4.3.16)). More briefly, each component of xA is at least v, and each component of $A\mathbf{y}^t$ is at most v.

It should be clear that the value, if it exists, is unique, representing a gain-floor for P_1 and a loss-ceiling for P_2. We shall prove (in a subsequent section) that it always exists.

Applying this definition of a saddle point to the game of matching pennies, we find that the value is 0, with optimal strategies (1/2, 1/2) for both players. In fact,

$$(1/2, 1/2) \begin{vmatrix} -1 & 1 \\ 1 & -1 \end{vmatrix} = (0, 0)$$

and

$$\begin{bmatrix} -1 & 1 \\ 1 & -1 \end{bmatrix} \begin{bmatrix} 1/2 \\ 1/2 \end{bmatrix} = \begin{bmatrix} 0 \\ 0 \end{bmatrix}$$

so that all the components of **x**A and A**y**' are 0. It is not too difficult to see that the optimal strategies are unique. In fact, suppose **x** is a strategy for P_1. We will have

$$\mathbf{x}A = (x_2-x_1, x_1-x_2)$$

and it is clear that one of the two components of **x**A will be negative unless $x_2 = x_1$. Thus, only (1/2, 1/2) will have the security level 0 for P_1. A similar argument shows the uniqueness of P_2's optimal strategy.

IV.3.3. Example. (Scissors, Stone, and Paper). In a well-known Oriental game, the two players simultaneously expose a hand, which, depending on its position, represents scissors (two open fingers), stone (a clenched fist), or paper (an open hand). There is a payoff of one unit according to the rule: scissors cuts paper, paper covers stone, and stone breaks scissors. If both players show the same, the game is a standoff.

The matrix for this game is

	St	Sc	Pa
St	0	1	-1
Sc	-1	0	1
Pa	1	0	-1

It is not difficult to guess (from the symmetry of the game) that the value must be 0, with optimal strategies (1/3, 1/3, 1/3) for both players. In fact, it is easy to see that

$$(1/3, 1/3, 1/3)A = (0, 0, 0)$$

and

$$A(1/3, 1/3, 1/3)^t = (0, 0, 0)^t$$

So that the strategies $\mathbf{x}^* = \mathbf{y}^* = (1/3, 1/3, 1/3)$ guarantee the value 0.

4. SOLUTION OF 2×2 GAMES

Let A be a 2×2 game matrix. To look for optimal strategies, we look first of all for a saddle point (in pure strategies). This is easy, as it reduces to checking whether P_1's gain-floor in pure strategies (the maximum of the row minima) is equal to P_2's loss-ceiling (the minimum of the column maxima). If this saddle point exists, there is no further problem: this gives us the optimal strategies (which are then pure strategies). Assume, then, that no such saddle point exists. This means that the optimal strategies (if they exist) must be mixed strategies in which each player uses both of his pure strategies with positive probability.

Conditions (4.3.13) can be restated:

$$a_{11}x_1 + a_{21}x_2 \geq v$$
$$a_{12}x_1 + a_{22}x_2 \geq v$$

Suppose, however, that strict inequality held in either one of these two conditions. Say, for instance,

$$a_{11}x_1 + a_{21}x_2 = v + \varepsilon$$

where $\varepsilon > 0$. It is not difficult to see that, in this case, P_2 should never use his first column. For, if $y_1 > 0$, then

$$\mathbf{x}A\mathbf{y}^t = (a_{11}x_1 + a_{21}x_2)y_1 + (a_{12}x_1 + a_{22}x_2)(1-y_1) \geq (v+\varepsilon)y_1 + vy_1$$

or

$$\mathbf{x}A\mathbf{y}^t \geq v + \varepsilon y_1$$

which contradicts the optimality of $\mathbf{y}$. (Heuristically, we are saying here that it cannot be optimal for P_2 to consider using a strategy that gives him an expected loss greater than v against P_1's optimal strategy.) We conclude that, if P_2 is to use both of his strategies with positive probability (i.e., a mixed strategy), we must have

4.4.1 $\qquad a_{11}x_1 + a_{21}x_2 = v$
4.4.2 $\qquad a_{12}x_1 + a_{22}x_2 = v$

Similarly, if P_1 is to use both of his strategies with positive probability,

4.4.3 $\quad a_{11}y_1 + a_{12}y_2 = v$
4.4.4 $\quad a_{21}y_1 + a_{22}y_2 = v$

This gives us four equations. We have additional "strategy" conditions

4.4.5 $\quad x_1 + x_2 = 1$
4.4.6 $\quad y_1 + y_2 = 1.$

This gives us a system of six equations in only five variables ($x_1, x_2, y_1, y_2,$ and v). It so happens, however, that one of the six equations is redundant, and so a solution will always exist. Moreover, the solution will be unique and will also satisfy the non-negativity constraints $\mathbf{x} \geq \mathbf{0}, \mathbf{y} \geq \mathbf{0}$ unless the game has a saddle point in pure strategies. In fact, it is not difficult to see that the solution to this system is

4.4.7 $\quad x_1 = (a_{22} - a_{21})/H$

4.4.8 $\quad x_2 = (a_{11} - a_{12})/H$

4.4.9 $\quad y_1 = (a_{22} - a_{12})/H$

4.4.10 $\quad y_2 = (a_{11} - a_{21})/H$

4.4.11 $\quad v = (a_{11}a_{22} - a_{21}a_{12})/H$

where

$$H = a_{11} + a_{22} - a_{21} - a_{12}$$

Unless the game has a saddle point in pure strategies, the numerators in equations (4.4.7)-(4.4.10) will all have the same sign as H, and so the non-negativity constraints are satisfied.

A simpler way of expressing these equations would be as follows: if the game has no saddle points, take the game matrix

$$A = \begin{bmatrix} a_{11} & a_{12} \\ a_{21} & a_{22} \end{bmatrix}$$

Form the *adjoint* matrix, which is obtained by interchanging the two entries on the main diagonal, and taking the negatives of the other two entries:

$$A^* = \begin{bmatrix} a_{22} & -a_{12} \\ -a_{21} & a_{11} \end{bmatrix}$$

Form now the two vectors JA^* and A^*J^t, where $J = (1, 1)$. (This corresponds to adding the rows, and the columns, respectively, of matrix A^*.)

$$JA^* = (a_{22} - a_{21}, a_{11} - a_{12})$$

$$A^*J^t = (a_{22} - a_{12}, a_{11} - a_{21})^t$$

The optimal strategies will now be *proportional* to JA^* and A^*J^t respectively; i.e., JA^* and A^*J^t must be multiplied by some constant to guarantee that the sum of their components is equal to 1.

We give the following examples of the use of these equations to solve games.

IV.4.1. Example. (Modified Matching Pennies). Consider the game of matching pennies, with the following modification: if P_2 fails to guess P_1's call correctly, then he must pay 2 units to P_1. If both P_1 and P_2 call H, then P_1 must pay P_2 3 units. If both P_1 and P_2 call T, then P_1 must pay P_2 1 unit.

The game matrix here is

$$A = \begin{bmatrix} -3 & 2 \\ 2 & -1 \end{bmatrix}$$

At first sight, this games seems to be fair (value 0). Applying equations (4.4.10) to (4.4.11), however, gives us

$$x_1 = 3/8$$
$$x_2 = 5/8$$
$$y_1 = 3/8$$
$$y_2 = 5/8$$
$$v = 1/8$$

revealing that the game is favorable to P_1 as it has positive value $v = 1/8$. It is easily checked that the strategy (3/8, 5/8) is optimal for both players, guaranteeing the value.

IV.4.2. Example. A secret service agent must carry a message from her headquarters to a friend's office. She can take either of two routes, one of them deserted, the other through the center of a city. Her opponent, a corrupt law-enforcement officer, wishes to prevent her from delivering the message and must guess which route to guard. If the corrupt officer catches the agent on the deserted road, he will kill her; if he catches her on the city street, he will merely be able to arrest her and delay her for some time. We assign to the agent a utility of +5 if she delivers the message, of –1 if she is arrested, and of –100 if she is killed.

The matrix for this game is

$$\begin{bmatrix} -1 & 5 \\ 5 & -100 \end{bmatrix}$$

Application of equations (6.4.7) to (6.4.11) to this game gives us the optimal strategies and value

$$\mathbf{x} = (35/37, 2/37)$$
$$\mathbf{y} = (35/37, 2/37)$$
$$v = -25/37$$

Note that, according to the solution, P_1 takes the safe route (first row) with very high probability. She must, however, take the riskier route with some small probability, if only to make her opponent guess.

5. $2 \times N$ AND $M \times 2$ GAMES

In case one player has two strategies, but the other has more than two, it is still possible to solve the game by applying the foregoing rules with some modification. The fact is that, in any $m \times 2$ or $2 \times n$ game, there will always be either a saddle point (in pure strategies) or a 2×2 submatrix that gives us the solution of the game.

IV.5.1. Example. As an example, consider the 3×2 game

0	3	2
4	2	3

Which has no row or column domination and no saddle points. Let us suppose that P_2 restricts himself to two of his three strategies. Depending on which strategy (column) he discards, we will obtain one of the three matrix games

$$A_1 = \begin{bmatrix} 0 & 3 \\ 4 & 2 \end{bmatrix}$$

$$A_2 = \begin{bmatrix} 0 & 2 \\ 4 & 3 \end{bmatrix}$$

$$A_3 = \begin{bmatrix} 3 & 2 \\ 2 & 3 \end{bmatrix}$$

With values

$$v_1 = 12/5, \quad v_2 = 3, \quad v_3 = 5/2$$

respectively.

Of these values, v_1 is the smallest. Hence, if P_2 is going to restrict himself to two strategies, he should choose those that give matrix A_1: the first two columns. Under these restrictions, the optimal strategies, obtained by applying rules (4.4.7)-(4.4.11) to the matrix A_1 (and assigning probability 0 to the third column), are

$$\mathbf{x} = (2/5, 3/5)$$

$$\mathbf{y} = (1/5, 4/5, 0)$$

It is now a simple matter to check that these are optimal strategies for the larger game A, guaranteeing the value. In fact, we calculate that

$$\mathbf{x}A = (12/5, 12/5, 13/5)$$

so that $\mathbf{x}$ guarantees that the payoff will be at least 12/5. Similarly,

$$A\mathbf{y}' = (12/5, 12/5)'$$

and the value of the game is indeed 12/5.

Thus, a general rule for solving a $2 \times n$ matrix game might go as follows:
(1) Check for a saddle point.
(2) If there are no saddle points, check to see whether any columns are dominated. Delete these.
(3) From the remaining columns, try all possible combinations of two columns (each such giving a 2×2 matrix game). Solve all of these games.
(4) Take the 2×2 game which has the smallest value. The optimal strategies for this game are optimal for the original game (Assign probability 0 to all the missing columns.)

For $m \times 2$ games, the procedure is the same, though in step (2) we look for row dominations; in step (3) we look for the 2×2 game with the largest value; in step (4) we assign 0 probability to the missing rows.

IV.5.2. Example. Consider the game

$$A = \begin{bmatrix} 3 & 2 & 3 & 1 & 5 \\ 2 & 4 & 1 & 6 & 0 \end{bmatrix}$$

In this game, we note:
(1) there are no saddle points;
(2) the first column is dominated (by the third). This leaves us with

$$A' = \begin{bmatrix} 2 & 3 & 1 & 5 \\ 4 & 1 & 6 & 0 \end{bmatrix}$$

(1) From this matrix, we can form $_4C_2 = 6$ combinations of 2 columns. Of the 6 possible 2×2 games, the one with smallest value corresponds to the third and fourth columns of A (second and third columns of A') with value 17/7. Applying (6.4.7)-(6.4.11), we obtain

$$\mathbf{x} = (5/7, 2/7)$$
$$\mathbf{y} = (0, 0, 5/7, 2/7, 0).$$

Again, we check by computing

$$\mathbf{x}A = (19/7, 18/7, 17/7, 17/7, 25/7)$$
$$A\mathbf{y}^t = (17/7, 17/7)^t.$$

So that the two strategies guarantee the value $v = 17/7$.

It is sometimes even possible, by using the symmetry of a matrix, to reduce larger games to the form of a 2×n or m×2 matrix.

IV.5.3. Example. The well-known military strategist, Colonel Blotto, has three divisions, which he must use to capture as many as possible of three mountain passes. He is faced by an opponent with two divisions. We shall assume, first, that divisions cannot be split, and, second, that a superiority of one division (or more) at a mountain pass will guarantee capture of that pass, along with any enemy divisions at that pass. Blotto scores one point for each pass he captures, and one point for each enemy division overpowered. He loses one point in each contrary instance.

We will designate each strategy as a triple denoting the number of strategies sent to each pass. Thus, 201 represents Blotto's strategy of sending two divisions to the first pass and one division to the third pass. If, for example, Blotto uses 201, while his opponent uses 002, we find that Blotto captures the first pass, scoring one point, but loses the third pass, and additionally has a division overpowered at that pass, giving him a net of −1 point. We obtain, in this manner, the 10×6 matrix

	200	020	002	110	101	011
300	3	0	0	1	1	-1
030	0	3	0	1	-1	1
003	0	0	3	-1	1	1
210	1	-1	1	2	2	0
201	1	1	-1	2	2	0
120	-1	1	1	2	0	2
021	1	1	-1	2	0	2
102	-1	1	1	0	2	2
012	1	-1	1	0	2	2
111	0	0	0	1	1	1

This game is much too large to solve by the methods which we have so far discussed. It is not difficult to see, however, that Blotto's strategies can be classified into three types, which can be called, respectively, 3, 21, and 11, according to the way his forces are divided, but regardless of destination. By symmetry, it is also possible to see that strategies of the same type should be used with equal probability. Similarly, his opponent's strategies can be classified into two types, called 2 and 11 respectively. Again, symmetry suggests that strategies of the same type should be used with equal probabilities.

We can think of this as a two-stage decision process for each player: first, each player decides the *type* of strategy he will use; later, the particular strategy. If Blotto decides to use type 21, and his opponent decides to use 11, and then strategies from each type are chosen with equal probabilities, we find that the expected payoff to Blotto is 4/3 points (2 points in 2/3 of the cases, and in 1/3 of the cases). In this manner, we obtain a reduced matrix

	2	1,1
3	1	1/3
2,1	1/3	4/3
1,1,1	0	1

In this matrix, we see that the second row dominates the third. We are left then with the 2×2 game

1	1/3
1/3	4/3

With solution $\mathbf{x} = (3/5, 2/5)$, $\mathbf{y} = (3/5, 2/5)$, $v = 11/5$. The 3×2 game will have optimal strategies $\mathbf{x} = (3/5, 2/5, 0)$, $\mathbf{y} = (3/5, 2/5)$, and same value. Now, remembering that strategies of the same type must have equal probabilities, and that the $\mathbf{x}$ and $\mathbf{y}$ above give us the total probabilities for all strategies of the same type, we obtain the optimal strategies

$$\mathbf{x^*} = (1/5, 1/5, 1/5, 1/15, 1/15, 1/15, 1/15, 1/15, 1/15, 0)$$

$$\mathbf{y^*} = (1/5, 1/5, 1/5, 2/15, 2/15, 2/15)$$

and value

$$v = 11/15$$

for the original 10×6 game.

The reader should note that the reduction here was possible only because of the complete symmetry existing among strategies of the same type. Generally, this type of thing is not possible.

PROBLEMS ON $m \times 2$ AND $2 \times n$ GAMES

1. Solve the following matrix games.

(a)
3	5
8	1

(b)
4	6	3
1	2	5

(c)
2	4	3
1	0	2

(d)
1	5	7
6	2	8

(e)
6	2	1
3	4	5

2. A game is played as follows. Player P_1 is given the aces of spades and diamonds, and the deuce of hearts. P_2 is given the aces of hearts and clubs, and the deuce of spades. Each then chooses a card (from his hand), and the two cards are shown simultaneously, with payoffs as follows. If both cards are of the same suit, then the ace beats the deuce. If the cards are of different suits and different denominations, then the black card beats the red. If they are of the same denomination, then the red card beats the black. If they are both black aces or both red aces, then there is a standoff. In any case, the player winning will win as many dollars as there are spots on his opponent's card. Find optimal strategies and the value of this game.

3. Colonel Blotto, commanding two divisions, must capture an enemy outpost defended by one division. To do this, he must have a numerical superiority of

one division when he attacks the outpost. It may be, however, that while Blotto attacks, his opponent simultaneously attacks Blotto's camp. If so, a superiority of one division is also sufficient to capture Blotto's camp. Blotto gains 1 unit (of payoff) if he captures the opponent's camp without simultaneously losing his own; he loses 1 unit if the opponent captures his camp. Otherwise the game is a standoff. What is Blotto's optimal strategy?

6. SOLUTIONS BY LINEAR PROGRAMMING

It was mentioned earlier that matrix games are related to linear programs. The connection is quite apparent, since the two *hemigames* (the problems faced by the two players in a game) are respectively a maximizing and a minimizing problem. In this section we shall formulate these two problems as linear programs, and show that the two programs are mutually dual.

As we know, an optimal strategy can be define as one which gives a security level equal to the value of the game. Since no higher security level is available for P_1, it follows that P_1's search for optimal strategies reduces to maximization of his security level. The strategy **x** gives a security level λ (or better) if every component of the vector **x**A is at least equal λ:

4.6.1 $$\sum_{i=1}^{m} a_{ij} x_i \geq \lambda \quad j = 1, 2, \ldots, n$$

Now, P_1 must maximize λ, subject to (4.6.1) and the additional constraint that **x** must be a strategy. We therefore obtain the linear program

4.6.2 Maximize λ

Subject to

4.6.3 $$-\sum_{i=1}^{m} a_{ij} x_i + \lambda \leq 0 \quad j = 1, 2, \ldots, n$$

4.6.4 $$\sum_{i=1}^{m} x_i = 1$$

4.6.5 $\quad\quad\quad\quad\quad\quad x_i \geq 0 \quad i = 1, 2, ..., m$

(Note that λ is unrestricted in the sense that it may take on negative values.)

In a similar way, P_2 must minimize his security level (loss-ceiling), μ. This gives us the program

4.6.6 $\quad\quad\quad\quad\quad\quad$ Minimize μ

$\quad\quad\quad\quad\quad\quad\quad\quad\quad\quad$ Subject to

4.6.7 $\quad\quad\quad\quad\quad -\sum_{j=1}^{n} a_{ij} y_j + \mu \geq 0 \quad i = 1, 2, ..., m$

4.6.8 $\quad\quad\quad\quad\quad\quad \sum_{j=1}^{n} y_j = 1$

4.6.9 $\quad\quad\quad\quad\quad\quad y_j \geq 0 \quad j = 1, 2, ..., n$

(Again, note that μ is an unrestricted variable.)

As with most linear programs, a question arises as to the duals. What, for instance, is the dual of (4.6.2) to (4.6.5)? If we go back to our study of duality in linear programs, (including the fact that an unrestricted variable corresponds to an equation, as we saw with transportation problems), we will see that the dual of (4.6.2) to (4.6.5) is precisely (4.6.6) to (4.6.9). In other words, the two hemigames are mutually dual programs. We saw in Chapter II that, in general, the dual of a linear program can be interpreted only in terms of "shadow prices," but we here have a case in which a linear program and its dual both have practical (and, in fact, very similar) interpretations.

We know, of course, that both programs (4.6.2) to (4.6.5) and (4.6.6) to (4.6.9) are feasible. We can therefore apply Theorem II.11.5 (the principal theorem of linear programming) to obtain a very important theorem.

IV.6.1. Theorem. Every matrix game has a value.

Theorem IV.6.1 is generally known as the minimax theorem, since it was originally given in the form

4.6.10 $$\min_{y \in Y} \max_{x \in X} xAy' = \max_{x \in X} \min_{y \in Y} xAy'$$

The fact that the two hemigames are mutually dual programs has one additional advantage: not only are the two security levels equal, but computation of P_1's optimal strategy (by using the simplex algorithm) will automatically give us P_2's optimal strategy as well.

IV.6.2. Example. Compute the optimal strategies and value for the matrix game

3	6	1	4
5	2	4	2
1	4	3	5

We write the maximizing problem in the form

Maximize λ
Subject to
$$3x_1 + 5x_2 + x_3 \geq \lambda$$
$$6x_1 + 2x_2 + 4x_3 \geq \lambda$$
$$x_1 + 4x_2 + 3x_3 \geq \lambda$$
$$4x_1 + 2x_2 + 5x_3 \geq \lambda$$
$$x_1 + x_2 + x_3 = 1$$
$$x_1, x_2, x_3 \geq 0$$

This gives us the tableau

	x_2	x_3	λ	
x_1				
-	-5	-1	1	$=-u_1$
3				
-	-2	-4	1	$=-u_2$
6				
-	-4	-3	1	$=-u_3$
1				
-	-2	-5	1*	$=-u_4$
4				
-	-1	-1	0	$=-1$
1				

This tableau is not in standard form for linear programs: λ is the objective function and should therefore be put among the basic variables (actually, on the bottom rows of the tableau). On the other hand, the 1 should be taken from among the basic variables and put at the top of the right-hand column. Unfortunately, the right-hand bottom entry is 0 and cannot be used as pivot, so we will need two pivots to accomplish this. We pivot on the starred entries to obtain the following sequence of tableaux:

x_1	x_2	x_3	u_4	
1	-3	4	-1	$=-u_1$
-2	0	1	-1	$=-u_2$
3	-2	2	-1	$=-u_3$
-4	-2	-5	1	$=-\lambda$
-1	-1	-1*	0	$=-1$

x_1	x_2	1	u_4	
-3	-7	4	-1	$=-u_1$
-3	-1	1	-1	$=-u_2$
1	-4	2	-1	$=-u_3$
1	3	-5	1	$=-\lambda$
1	1	-1	0	$=-x_3$

We now interchange the third and fourth columns, and also the fourth and fifth rows, to place the 1 and the λ in the desired positions. We also change the sign in the row corresponding to λ:

	x_1	x_2	u_4	1	
	-3	-7	-1	4	$=-u_1$
	-3	-1	-1	1	$=-u_2$
	1	-4	-1*	2	$=-u_3$
	1	1	0	-1	$=-x_3$
	-1	-3	-1	5	$=\lambda$

We proceed now according to the usual rules until the solution of the program is obtained:

	x_1	x_2	u_3	1	
	-4	-3	-1*	2	$=-u_1$
	-4	3	-1	-1	$=-u_2$
	-1	4	-1	-2	$=-u_4$
	1	1	0	-1	$=-x_3$
	-2	1	-1	3	$=\lambda$

	x_1	x_2	u_1	1	
	4	3	-1	-2	$=-u_3$
	0	6*	-1	-3	$=-u_2$
	3	7	-1	-4	$=-u_4$
	1	1	0	-1	$=-x_3$
	2	4	-1	1	$=\lambda$

This gives us a basic feasible point. We now proceed as always in Stage II.

	x_1	u_2	u_1	1	
	4*	-1/2	-1/2	-1/2	$=-u_3$
	0	1/6*	1/6	-1/2	$=-x_2$
	3	-7/6	1/6	-1/2	$=-u_4$
	1	-1/6	1/6	-1/2	$=-x_3$
	2	-2/3	-1/3	3	$=\lambda$

	u_3	u_2	u_1	1	
v_1	1/4	-1/8	-1/8	-1/8	$=-x_1$
v_2	0	1/6*	1/6	-1/2	$=-x_2$
y_4	-3/4	-19/24	13/24	-1/8	$=-u_4$
v_3	-1/4	-1/24	7/24	-3/8	$=-x_3$
-1	-1/2	-5/12	-1/12	13/4	$=\lambda$
	$=y_3$	$=y_2$	$=y_1$	$=\mu$	

This last tableau gives us the solution of the game: P_1's optimal strategy is

$$\mathbf{x} = (1/8, 1/2, 3/8)$$

The dual variables, which were omitted in the intermediate steps, tell us that P_2's optimal strategy is

$$\mathbf{y} = (1/12, 5/12, 1/2, 0)$$

The value of the game is 13/4. This may be checked directly by forming the products $\mathbf{xA}$ and $\mathbf{Ay}^t$.

While it is normal to reduce a matrix game top a linear program (for purposes of solution) it is, conversely, possible to reduce a linear program to a matrix game. To show how this is done, we give first a definition.

IV.6.3. Definition. A square matrix, A, is *skew-symmetric* if $A = -A^t$.

Examples of skew-symmetric matrices are

$$\begin{bmatrix} 0 & 1 \\ -1 & 0 \end{bmatrix} \quad \begin{bmatrix} 0 & 1 & -3 \\ -1 & 0 & 2 \\ 3 & -2 & 0 \end{bmatrix} \quad \begin{bmatrix} 0 & 4 & 1 \\ -4 & 0 & 2 \\ -1 & -2 & 0 \end{bmatrix}$$

A game with skew-symmetric matrix is said to be *symmetric*, and it is not difficult to see that such a game must have value 0.

Take, now, the pair of dual linear programs

4.6.11   Maximize xc^t
          Subject to

4.6.12 $xA \leq b$
4.6.13 $x \geq 0$

and

4.6.14   Minimize yb^t
          Subject to
4.6.15 $yA^t \geq c$
4.6.16 $y \geq 0$

and consider the $(m+n+1) \times (m+n+1)$ matrix game

$$M = \begin{bmatrix} \Theta & A^t & -b^t \\ -A & \Theta & c^t \\ b & -c & 0 \end{bmatrix}$$

where the Θ's are zero matrices of the correct size, and A, **b, c**, are as just given. It is easy to see that the matrix M is skew-symmetric and so must, as a game, have value 0.

Let

$$\mathbf{p} = (p_1, \ldots, p_n; p_{n+1}, \ldots, p_{m+n}; p_{m+n+1})$$

be an optimal strategy (for player P_1) in this game, such that $p_{m+n+1} > 0$, and define **x** and **y** by

4.6.17 $\qquad x_i = p_{n+i}/p_{m+n+1} \quad$ for $i = 1, \ldots, m$

4.6.18 $\qquad y_j = p_j/p_{m+n+1} \quad$ for $j = 1, \ldots, n.$

Then, it is not difficult to see that **x** and **y** solve the two programs (4.6.11) to (4.6.13) and (4.6.14) to (4.6.16) respectively.

If $p_{m+n+1} > 0$, it is of course impossible to do this. However, it may be shown that, if both programs are feasible, then there will always be at least one optimal strategy, **p**, for the game M, with $p_{m+n+1} > 0$.

PROBLEMS ON LINEAR PROGRAMMING FOR MATRIX GAMES

1. Solve the following matrix games:

(a)

3	5	1	2
4	6	0	1
2	1	5	3

(b)

3	5	2
4	2	6
1	6	3

(c)

1	3	1	5
2	0	4	3
5	2	3	0

(d)

3	0	1
1	5	2
2	3	0

(e)

0	1	-2
-1	0	3
2	-3	0

2. Show that, if a game is symmetric (i.e. its matrix is skew-symmetric), then the value must be zero. [Hint: using the fact that for each i and j, $a_{ij} = -a_{ji}$, compute the product $\mathbf{x}A\mathbf{x}'$, where $\mathbf{x}$ can be any strategy.]

3. Two duelists, each holding a pistol with one shot, advance towards each other. After taking n steps towards each other (if neither has been shot before this) they will be next to each other. Either one can shoot after any number k of steps, $1 \leq k \leq n$, and so each player has n strategies.

   Assume that, if one of the duelists shoots after k steps, there is a probability k/n that he will hit his opponent. In such case, the duel ends immediately. Assume also that the duel is "silent", i.e. a player does not know whether his opponent has yet shot, unless he himself is hit.

   Assign a payoff of +1 if P_1 hits his opponent without being himself hit, and −1 in the contrary instance. The game is a standoff (payoff 0) if neither one is hit, or if they are both hit (simultaneously).

   (a) Compute the game matrix.
   (b) Solve the game for $n = 2, 3, 4$, and 5.

4. Two players are given respectively the ace, deuce and trey of spades, and the ace, deuce, and trey of hearts. The ace, deuce, and trey of diamonds are placed, face up, on the board. Each player then places one of his cards (face down) next to one of the diamonds. The cards are then turned, and the player with the higher card captures the corresponding diamond card. (The ace is considered low.) The payoff is the difference in the number of spots on the diamond cards captured.

   Find optimal strategies for this game.

5. Modify problem 4 by assuming that the cards are the deuce, trey, and four of each suit.
6. Modify problem 3, pageXXX, to the case where Blotto has three divisions, and his opponent two divisions. Find optimal strategies and the value.

7. SOLUTION OF GAMES BY FICTITIOUS PLAY

Let us consider the following hypothetical situation: a pair of players, perhaps ignorant of game theory, play a game a large number of times. While ignorant of game theory, they are, nevertheless, statistically inclined, and so each keeps track of the strategies used by his opponent over the previous trials of the game. Each then proceeds to play according to the rather naïve rule of choosing that pure strategy which is optimal against his opponent's observed mixture of pure strategies.

To express this mathematically, let us suppose that, during the first k trials of the game, P_2 has used his jth strategy Y^k_j times. Then, on the $(k+1)$th trial, P_1 uses that row, i, which maximizes the sum

4.7.1 $$\sum_j a_{ij} Y^k_j .$$

Similarly, if P_1 has used his ith strategy X^k_i times during the first k trials, then P_2 should, on the $(k+1)$th trial, use that strategy, j, which minimizes the sum

4.7.2 $$\sum_i a_{ij} X^k_i .$$

There are two things to be said about this procedure. The first is that it is easily carried out since it is merely a question of keeping track of the running totals (4.7.1) and (4.7.2). The second is that it is wonderfully naïve, as it seems to suggest that each of the two players believes he is playing against an irrational opponent (say, nature). And yet, the fact is that, naïve as the procedure seems, it nevertheless solves the game.

More precisely, suppose that two sequences of vectors, $\mathbf{X}^k$ and $\mathbf{Y}^k$, are defined as follows:
$$X^0_i = 0 \quad \text{for all } i = 1, 2, ..., m$$
$$Y^0_j = 0 \quad \text{for all } j = 1, 2, ..., n$$

Assuming X^{k-1} and Y^{k-1} have been calculated, let $h(k)$ be that value of i which maximizes $\sum a_{ij} Y^k_j$. Let $l(k)$ be that value of j which minimizes $\sum a_{ij} X^k_i$. Then

$$X^k_{h(k)} = X^{k-1}_{h(k)} + 1.$$
$$X^k_i = X^{k-1}_i \quad \text{if } i \neq h(k).$$

$$Y^k_{l(k)} = Y^{k-1}_{l(k)} + 1.$$
$$Y^k_j = Y^{k-1}_j \quad \text{if } j \neq l(k).$$

Note: in case of ties (for maximizing i or minimizing j) any one of those tied for the maximum of minimum may be chosen.

We then have the following theorem:

IV.7.1. Theorem. Let the vectors $\mathbf{X}^k$ and $\mathbf{Y}^k$ be generated as just explained, and define

4.7.3 $\qquad\qquad\qquad\mathbf{x}^k = \mathbf{X}^k/k$
4.7.4 $\qquad\qquad\qquad\mathbf{y}^k = \mathbf{Y}^k/k$

Then $\mathbf{x}^k$ and $\mathbf{y}^k$ are mixed strategies, and, moreover, $u(\mathbf{x}^k)$ and $w(\mathbf{y}^k)$ both tend to v (the value of the game) as k increases.

Theorem IV.7.1, which is too difficult to prove here, was originally conjectured by G. W. Brown and later proved by J. Robinson. It states that, as the number of repetitions increases, the mixed strategies generated by this method will be nearly optimal. Thus, by taking sufficiently many iterations, the value and optimal strategies may be approximated as closely as desired. Of course, the number of iterations (value of k) required for a good approximation may be quite large, but then each iteration is much simpler than a pivot step.

This technique is generally known as the method of *fictitious play*.

To give an example of this, consider the game with matrix

$$A = \begin{array}{|cccc|} \hline 3 & 1 & 6 & 2 \\ 5 & 8 & 4 & 2 \\ 2 & 3 & 1 & 6 \\ 1 & 5 & 3 & 4 \\ \hline \end{array}$$

The initial choice may be made arbitrarily; perhaps the best way is to choose that pure strategy with best security level, so on that basis P_1 should choose row 2 while

P_2 should choose column 1. Thus $h(1) = 2$, $l(1) = 1$. Against P_1's choice of row 2, column 4 is best, so $l(2) = 4$, and against P_2's choice of column 1, row 2 is best, so $h(2) = 2$. Proceeding in this manner, we find that, on the first 20 trials, P_1's choices are

$$2,2,3,3,3,1,1,1,1,1,2,2,2,2,2,2,2,3,3,3$$

while P_2's choices are

$$1,4,4,3,3,3,3,1,1,1,2,4,4,4,4,4,4,4,4,4$$

which gives us the "empirical" strategies

$$x^{20} = (0.25, 0.45, 0.30, 0)$$

$$y^{20} = (0.20, 0.05, 0.20, 0.55).$$

It is now easy to calculate that

$$xA = (3.6, 4.75, 3.6, 3.2)$$

and

$$Ay^t = (2.95, 3.3, 4.05, 3.25)$$

So that the security levels are $u(x) = 3.2$ and $w(y) = 4.05$. The "gap" $w(y)-u(x) = 0.85$, which is still larger than we might want. With 50 iterations, we would find

$$x^{50} = (0.32, 0.40, 0.28, 0)$$

$$y^{50} = (0.32, 0.02, 0.28, 0.38)$$

with the security levels $u(x) = 3.12$ and $w(y) = 3.64$, respectively. The gap is now 0.52, which is somewhat better. The exact optimal strategies (which we obtained by use of the simplex algorithm) are

$$x = (0.25, 0.39, 0.36, 0)$$

$$y^{50} = (0.3,9\ 0, 0.29, 0.43)$$

with the value $v = 3.43$, so we see that the approximation, while not nearly perfect, is really not too bad.

This method is especially useful for the solution of very degenerate linear programs (reduced to games as explained in Section 5 of this chapter), if we are willing to settle for approximate solutions. The reason is that degeneracy in linear programs generally causes some difficulty in application of the simplex algorithm, while it has no effect in the method if fictitious play.

PROBLEMS ON SOLUTION OF GAMES BY FICTITIOUS PLAY

Solve the following games by fictitious play:

1.

3	5	1	2
2	6	3	4
6	1	2	0
5	3	4	1

2.

6	5	1	2	6
3	1	4	3	3
2	5	8	6	1
4	2	3	5	3

3.

1	6	1	4	2
3	0	4	1	8
1	2	5	3	4
2	1	6	1	0

4. Show that the linear program

$$\text{Maximize} \quad w = \mathbf{cx}'$$

$$\text{Subject to} \quad \mathbf{Ax}' \leq \mathbf{b}'$$
$$\mathbf{x} \geq \mathbf{0}$$

is equivalent to finding optimal strategies for P_2 in the game with matrix $\mathbf{A}' = (a_{ij}')$, where

$$a_{ij}' = a_{ij}/b_i c_j$$

and that the maximum value of w is equal to the reciprocal of the value of the game.

5. Use the results of problem 4, above, and the method of fictitious play to obtain an approximate solution to the linear program

Maximize
$$w = 5x_1 + 3x_2 + x_3 + 4x_4 + 2x_5$$
Subject to
$$3x_1 + 2x_2 + 4x_3 + x_4 + x_5 \leq 6$$
$$x_1 + 3x_2 + x_3 + 2x_4 + 3x_5 \leq 8$$
$$2x_1 + 4x_2 + 5x_3 + 2x_4 + x_5 \leq 12$$
$$x_1 + 3x_2 + 2x_3 + 4x_4 + x_5 \leq 8$$
$$x_1, x_2, x_3, x_4, x_5 \geq 0$$

8. THE VON NEUMANN MODEL OF AN EXPANDING ECONOMY

We give here, as an application of game-theoretical analysis (thug not of game theory itself), a model, developed by J. von Neumann, of an expanding economy.

The von Neumann model assumes that the economy produces n fundamental goods by means of m fundamental processes. The goods produced are in turn needed as inputs for the processes, so that the production at a future time is constrained by production in the present. More exactly, the inputs in period $k+1$ can be no greater than the outputs in period k.

The processes can be characterized by two m×n matrices, $A = (a_{ij})$ and $B = (b_{ij})$. The matrix A is the *input* matrix: the entry a_{ij} is the amount of product j needed as an input to operate the ith process at unit intensity. In turn, B is the *output* matrix: b_{ij} is the amount of product j produced in operating the ith process at unit intensity.

To explain this by an example, consider the case of an undeveloped country which is trying to "enter the twentieth century" by expanding its steel industry. To do this, it must construct additional capacity (i.e. more steel mills). This is complicated, however, because the capacity itself must be built of steel; thus, construction of the new capacity is bounded by the old capacity.

To simplify this example, let us suppose that labor, raw materials, and certain other industrial products (e.g. cement) are in plentiful supply. We shall also assume that none of the steel produced goes into the production of automobiles or other consumer goods (this would cause a permanent "leak" in our model). We have, then, an economy with two fundamental product: steel and steel mills, and two fundamental industries: steel and construction.

Assume that 1 ton of steel is necessary to construct a mill of 12-ton capacity. Assume also that 1/6 ton of steel is consumed in the production of each new ton of steel. We obtain, then, the following input-output matrices:

		Capacity	Steel
A =	Construction	0	1
	Siderurgical	1	1/6

		Capacity	Steel
B =	Construction	12	0
	Siderurgical	1	1

Note that in the second process (the siderurgical industry) the capacity that goes in as input comes out once again as an output of the process: this represents the fact that the steel mill is a capital good: it is not consumed in the production of steel, and will still be there after the steel has been forged.

Let us see how this economy can expand in practice. The principal question is whether such an economy can expand at a steady rate, without the difficulties (such as unemployment and shortage of goods) that often characterize overly quick growth. Suppose the country starts the (first) period with 300 T. of steel and the capacity for producing 600 T. In the first time period, 100T. will be used for production of new steel, and the remaining 200 T. will be used to construct new mills with a capacity of 2400 T. At the end of the first period, the country will then have an inventory of 600 T. of steel, and the capacity to produce 3000 T. In the second period, then, 500 T. will be used to produce new steel, and the other 100 T. for construction of new mills (with 1200 T. capacity) so that at the end of the second period, there are 3000 T. of steel in inventory and capacity to produce 4200

T. In the third period, 700 T. will be used in production and 2300 for construction, and at the end of the period there are 4200 T. in inventory, and steel mills with a total capacity or 31,800 T. But now there is not enough steel available to prime the mills (as this would require 5300 T.) Thus in the fourth period some capacity will go unused. There is now a danger of severe unemployment in the construction industry since there is no point in any new mills.

Suppose on the other hand that the country starts with 200 T. in inventory and the capacity to produce 600 T. Then in the first period 100 T. will be used to produce more steel, and the other 100 T. for construction. At the end of the first period, the country has 600 T. in inventory, and the capacity to produce 1800 T. In the second period, 300 T. will be used for production, and 300 T. for construction, giving a total of 1800 T. in inventory, and a capacity for 5400 T., at the end of the second period. We see that, in this case, the country's assets triple in each time period: the expansion can be continued indefinitely in this manner (until, of course a shortage of some other goods, currently in surplus, begins to make itself felt). We say then that the economy is *expanding in equilibrium with an expansion factor* 3.

In the general case, suppose that the ith process is being carried out at intensity x_i. This will mean that it uses $a_{ij}x_i$ units of the jth good. We can represent the operation of the entire economy by an *intensity vector* $\mathbf{x} = (x_i, \ldots, x_m)$. In this case the total amount of goods j needed as inputs is given by

$$\sum_{i=1}^{m} a_{ij} x_i$$

and so the total input of the economy can be expressed by the vector $\mathbf{xA}$. In a similar manner, the total amount of goods j obtained as outputs will be

$$\sum_{i=1}^{m} b_{ij} x_i$$

and so the total output is represented by the vector $\mathbf{xB}$.

Now the input for one cycle must be taken from the output of the previous cycle. As we are interested in a steady expansion of the economy, we want the intensity for each cycle to be some scalar, α, times the intensity in the previous cycle. (In the foregoing example, $\alpha = 3$.) If the intensity for the last cycle was $\mathbf{x}$, then the intensity for the current cycle must be αx. The feasibility constraint can then be represented by the inequality

$$(\alpha x)A \leq xB$$

or, equivalently,

4.8.1
$$x(B - \alpha A) \geq 0$$

A related result is that of finding "natural" prices for the economy. In fact, in the foregoing example, suppose that a ton of steel costs $150, while the capacity to produce this steel cost $50 (i.e., a steel mill with a capacity of t tons is valued at $50t.) Then operation of the construction industry at unit intensity requires an investment of $150 and produces $600. (Again, we are disregarding the costs of labor and other raw materials for the purpose of simplification; these will be introduced later on.) On the other hand, operation of the steel industry at unit capacity will require an investment of $75 ($50 for capacity plus $25 for steel) and give a return of $200. We see that the investment in the steel industry is not quite tripled in one time period, whereas investment in the construction industry is quadrupled in one period. This will cause greater investment in construction, while steel will be neglected – which will also cause severe economic dislocations.

Suppose, on the contrary, that a ton of steel costs $180, while a 1-tom production capacity is valued at $45. In this case, unit operation of the construction industry requires a $180 investment and gives a return of $540, while unit operation of the steel industry requires a $75 investment and gives a return of $225. In this case, both industries triple the invested money, so there is again an equilibrium: neither is more profitable than the other. This factor of 3 is also the "natural" interest rate: if the interest rate were lower, it would be "forced up" by people investing in the industries.

In the general case, we can let y_j be the price of one unit of the jth good. The vector $\mathbf{y} = (y_1, y_2, \ldots, y_n)$ is then the *price* vector. Now, the cost of operating the ith industry at unit intensity is given by

$$\sum_{j=1}^{n} a_{ij} y_j$$

and the return from this industry is

$$\sum_{j=1}^{n} b_{ij} y_j$$

The two vectors, Ay^t and By^t, represents the investments and returns from the several industries.

As was mentioned earlier, no industry can give a greater return than the prevailing interest rate (otherwise this rate will be forced up by people investing to seek the higher profit from such an industry). Thus if β is the interest rate, we must have

$$By^t \leq \beta Ay^t$$

or, equivalently,

4.8.2 $$(B-\beta A) y^t \leq 0$$

We have considered an example, and seen that the equilibrium expansion and interest rates are equal. The questions arise whether such equilibria will exist and whether, at equilibrium, the two rates, α and β, are always equal. Intuitively, this seems so, since an expansion rate α will normally "pull" β up to the same level or higher, i.e., $\alpha \leq \beta$. At the same time, investments will not be made unless the return is at least equal to the interest rate, i.e., $\alpha \geq \beta$.

We have two conditions, (4.8.1) and (4.8.2), concerning x, y, α, and β. Certain other conditions can also be obtained. For one, it is a known economic fact that surplus goods become worthless. The jth good is in surplus if more is produced during a cycle than is needed during the following cycle, i.e., if

$$(xB)_j > \alpha(xA)_j$$

Thus, we conclude that we must have $y_j = 0$ whenever $x(B-\alpha A)_j > 0$. This gives us the condition

4.8.3 $$x (B-\beta A) y^t = 0$$

An investment is unprofitable if its return is smaller than the going interest rate, β. It is natural that, in equilibrium, unprofitable investments are not made. Thus, we find that, if

$$(\mathbf{By}')_i < \beta(\mathbf{Ay}')_i$$

then the ith industry will not be operated: $x_i = 0$, which gives us

4.8.4 $\qquad \mathbf{x}(\mathbf{B}-\alpha\mathbf{A})\mathbf{y}' = 0.$

We shall put one final condition on $\mathbf{x}$ and $\mathbf{y}$. This is that the goods produced, at the going prices, must have some value. But the total value of goods produced is $\mathbf{xBy}'$. Thus

4.8.5 $\qquad \mathbf{xBy}' > 0$

We have obtained five conditions for equilibrium, enough to prove the equality of the expansion and interest rates. In fact, (4.8.3) and (4.8.4) can be restated in the form

$$\mathbf{xBy}' = \alpha\mathbf{xAy}'$$

$$\mathbf{xBy}' = \beta\mathbf{xAy}'$$

Thus

$$\beta\mathbf{xAy}' = \alpha\mathbf{xAy}'$$

But by (4.8.5), $\mathbf{xBy}'$ is not 0; hence $\mathbf{xAy}'$ is not zero, and we conclude that

4.8.6 $\qquad \alpha = \beta.$

We see, then, that interest and expansion rates must be equal. Thus (4.8.2) may be restated in the form

4.8.7 $\qquad (\mathbf{B}-\alpha\mathbf{A})\mathbf{y}' \leq 0.$

Consider now the vectors $\mathbf{x}$ and $\mathbf{y}$. Since these are necessarily non-negative, and by (4.8.5), neither one can be identically zero, each can be expressed as a positive scalar times a mixed-strategy vector. In our analysis, it is only the relative intensities of the industries and the relative prices of the goods that matter. It follows, then, that we may assume $\mathbf{x}$ and $\mathbf{y}$ are mixed-strategy vectors. The two constraints (4.8.1) and (4.8.7) mean that the matrix $\mathbf{B}-\alpha\mathbf{A}$ (thought of as a matrix game) has value

4.8.8 $$v(B-\alpha A) = 0$$

and that **x** and **y** are optimal strategies. Note, now, that (4.8.8) will guarantee the existence of **x** and **y** satisfying (4.8.1) and (4.8.7). Thus the question of the existence of an equilibrium reduces to the two conditions:

$$v(B-\alpha A) = 0$$

$$xBy' > 0$$

where **x** and **y** are optimal strategies for the game $B-\alpha A$.

9. EXISTENCE OF AN EQUILIBRIUM EXPANSION RATE

We consider next the question of the existence of an α such that (4.8.8) holds. So far, we have placed no conditions on the matrices A and B, other than the obvious one of non-negativity. We want our process, however, to be realistic. This means that we cannot produce something out of nothing. Every process must use some goods. Therefore, each row in the matrix A has at least one positive entry. This will mean that

4.9.1 $$v(-A) < 0$$

Another condition for realism is that every good must be produced by some process. Thus each column of B must have at least one positive entry. This means

4.9.2 $$v(B) > 0$$

Consider now the matrix game $B-\alpha A$. As A increases (from 0 to ∞) it is clear that the entries in $B-\alpha A$ will either decrease or remain the same. In this way the value $v(B-\alpha A)$ will decrease steadily or at least never increase. It is, moreover, a continuous function of α. For small values of α, the term αA is negligible, so $v(B-\alpha A)$ will be close to $v(B)$, which is positive. When α is large, the term B becomes negligible (compared to αA), and so $v(B-\alpha A)$ will be close to $v(-\alpha A)$, which is negative. Given the continuity of v, we see that, for some value of α, (6.88) must hold.

In our previous example, the matrix

$$B-\alpha A = \begin{array}{|cc|} \hline 12 & -\alpha \\ & \\ 1-\alpha & 1-\alpha/6 \\ \hline \end{array}$$

has no saddle points for $\alpha \geq 0$. Its value, computed by the method of Section 5, above is

$$v(B-\alpha A) = (12 - \alpha - \alpha^2) / (12 + 11\alpha/6)$$

FIGURE IV.9.1. The value of the game, $B-\alpha A$, as a function of α.

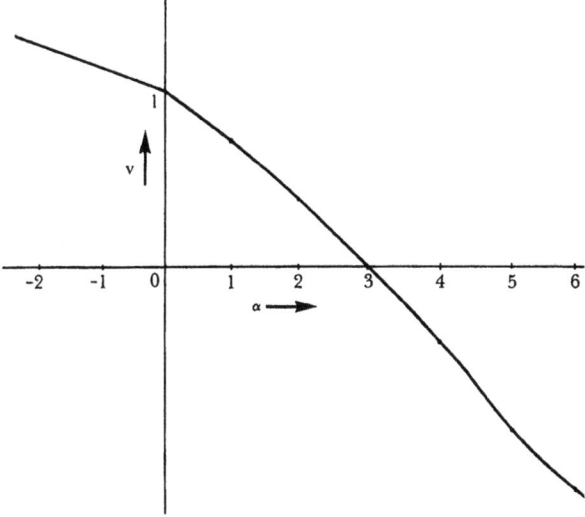

(see the graph in Figure IV.9.1). To make this value vanish, it suffices to let the numerator vanish, giving us the equation

$$12 - \alpha - \alpha^2 = 0$$

This equation has the solutions $\alpha = 3$ and $\alpha = -4$. The negative solution is extraneous (i.e., it corresponds to a value of α for which the game has a saddle point, so that the formula used for computing the value here does not apply). We see then that this economy can have an equilibrium only at expansion rate 3. Then

$$B-3A = \begin{bmatrix} 12 & -3 \\ -2 & 1/2 \end{bmatrix}$$

Optimal strategies may be computed for this game, and it will be seen that these are proportional to the equilibrium intensity and price vectors obtained above.

One final question remains, and that is whether the optimal strategies for $B-\alpha A$ will satisfy (6.8.5). It can be shown (though we will not give the proof here) that one of two cases must hold. One possibility is that there is only one number α such that (4.8.8) holds. If so, then there are always optimal strategies satisfying (4.8.5). The other possibility is that there may be more than one α for which (4.8.8) holds; i.e., there may be numbers α' and α'', such that $\alpha' < \alpha''$, with

$$v(B-\alpha'A) = v(B-\alpha''A) = 0.$$

In this case, if α is any number between α' and α'', then it is not difficult to see that α also satisfies (4.8.8), since v either decreases or remains constant as α increases. It turns out, however, that most of these numbers α correspond to cases in which nothing of value is produced: i.e. (4.8.5) cannot be satisfied. There will always be, nevertheless, some values of α (finitely many of them) that satisfy both (4.8.8) and (4.8.5). These include both the minimal and the maximal values of α which satisfy (4.8.8). An example will perhaps make this more clear.

IV.9.1. Example. Consider a two-product economy with the products "wheat" and "diamonds," produced respectively by the processes "agriculture" and "jewelry." The input-output matrices are:

$$A = \begin{array}{c} \\ \text{Agriculture} \\ \text{Jewelry} \end{array} \begin{array}{cc} \text{Wheat} & \text{Diamonds} \\ \begin{bmatrix} 10 & 0 \\ 5 & 10 \end{bmatrix} \end{array}$$

$$B = \begin{array}{c} \text{Agriculture} \\ \\ \text{Jewelry} \end{array} \begin{array}{|cc|} \hline \text{Wheat} & \text{Diamonds} \\ 40 & 0 \\ \\ 0 & 20 \\ \hline \end{array}$$

We have, then,

$$B - \alpha A = \begin{array}{|cc|} \hline 40-10\alpha & 0 \\ \\ -5\alpha & 20-10\alpha \\ \hline \end{array}$$

Suppose we set $\alpha = 4$. For this factor,

$$B - 4A = \begin{array}{|cc|} \hline 0 & 0 \\ \\ -20 & -20 \\ \hline \end{array}$$

It is easy to see that this game has value 0. P_1's optimal strategy is $\mathbf{x} = (1,0)$, but any strategy will be optimal for P_2. At this expansion rate, we see that only wheat is produced; the jewelry industry must be neglected since it simply cannot expand so quickly. The prices are irrelevant since there is, in effect, only one commodity. The only constraint, now, is that

$$\mathbf{x}B\mathbf{y}' = 40\, y_1 > 0$$

so that y_1 must be positive if (4.8.5) is to hold. In other words: wheat must be worth something in this economy; it does not matter what diamonds are worth since they are not being produced.

Suppose, next, that $\alpha = 2$. We will have

$$B-2A = \begin{bmatrix} 20 & 0 \\ -10 & 0 \end{bmatrix}$$

Again, we have a game with value 0. A strategy **x** will be optimal for P_1 if (and only if) $x_1 \geq 1/3$. P_2's unique optimal strategy is **y** = (0,1). In this case, wheat becomes worthless. (In practice, what would happen is that diamonds would become very expensive as compared to wheat.) The reason for this is that wheat production can expand more rapidly than the going interest rate $\alpha = 2$, and so it becomes a surplus good whose production would get out of hand if it had any value. To satisfy (4.8.5), we must have an **x** satisfying $x_1 \geq 1/3$ and $x_2 > 0$.

Suppose, next, we take some α between 2 and 4, say $\alpha = 3$. We have

$$B-3A = \begin{bmatrix} 10 & 0 \\ -15 & -10 \end{bmatrix}$$

For this game, the value is still 0, but the unique optimal strategies are **x** = (1,0) and **y** = (0,1). But this would mean that only wheat, which is worthless, will be produced; i.e., (6.8.5) does not hold. The trouble here is that diamonds, unable to expand so quickly, will not be produced; wheat can expand this quickly, but there will always be a surplus. This will be true for any α such that $2 < \alpha < 4$.

We see from this example that a new difficulty arises in such cases. The possible expansion rate is not unique; moreover, a product which has value at some expansion rate becomes worthless at another rate. What actually happens is that the economy tends to split up into independent sub-economies. The thing to remember, however, is that these sub-economies can never exist in a vacuum: each one must pay wages to its workers, who proceed to link up the economy again by their demands. If, in the foregoing example, we can get the agricultural workers to buy some diamonds (in the form, say, of jewelry for themselves of for their spouses) the economy will be linked together again.

Let us introduce, therefore, the "wages" matrix, $W = (w_{ij})$. The entry w_{ij} means that, if process i is operating at unit intensity, then its workers must be paid an amount w_{ij} of good j. (We are assuming that workers are paid in goods rather than in money. In reality the situation is, of course, more complicated, since the workers are paid in money and then decide what to buy with it, but in effect the two procedures are not too dissimilar.)

Suppose we have a wage matrix

$$W = \begin{bmatrix} 2 & 2 \\ 1 & 1 \end{bmatrix}$$

Which means that for the agriculture industry to operate at unit intensity, it must pay its workers 2 units of wheat and 2 units of diamonds. The jewelry industry pays only half as much, perhaps because it requires less labor.

It now turns out that the required input is no longer the matrix A, but the sum A+W. In other words, the entrepreneur needs the commodities, both as inputs and capital goods, and to pay wages. We therefore need a factor α such that

4.9.3 $\qquad v(B - \alpha A - \alpha W) = 0$

Expansion will then be less rapid (but more realistic).

For our example, we have

$$B - \alpha A - \alpha W = \begin{bmatrix} 40-12\alpha & -2\alpha \\ -6\alpha & 20-11\alpha \end{bmatrix}$$

It is not too difficult to prove that there is no value of α for which the game would have a saddle point and value zero. We solve for α under the assumption that the game does not have a saddle point. In this case, application of (4.4.11) leads us to the equation

$$120\alpha^2 - 680\alpha + 800 = 0$$

which gives the solutions $\alpha = 4$ and $\alpha = 5/3$. The value $\alpha = 4$ gives us the matrix

$$B-4A-4W = \begin{bmatrix} -8 & -8 \\ -24 & -24 \end{bmatrix}$$

which clearly does not have value zero (i.e., $\alpha = 4$ is an extraneous root). The other solution, $\alpha = 5/3$, gives us

$$B-5(A+W)/3 = \begin{bmatrix} 20 & -10/3 \\ -10 & 5/3 \end{bmatrix}$$

which has the desired value and unique optimal strategies

$$\mathbf{x} = (1/3, 2/3)$$

$$\mathbf{y} = (1/7, 6/7)$$

Note how the price of diamonds is considerably greater than the price of wheat. This is due, of course, to many things, not least of which is the relative scarcity of diamonds (they cannot be produced as quickly).

We see here how the influence of workers' demands links up the economy, giving a unique expansion factor and price system. It can be roved (though we will not do so here) that the system will have a unique expansion factor if all the elements in the matrix A+B+W are positive. Heuristically, this means that each process should somehow be related to each product, either using it as an input or producing it as an output, or else obtaining it to pay as wages t its employees. (This condition is sufficient, but not necessary, for the uniqueness of the expansion rate.)

PROBLEMS ON THE VON NEUMANN MODEL

1. Fine the intensity and price vectors at equilibrium in the economies represented by the following pairs of matrices:

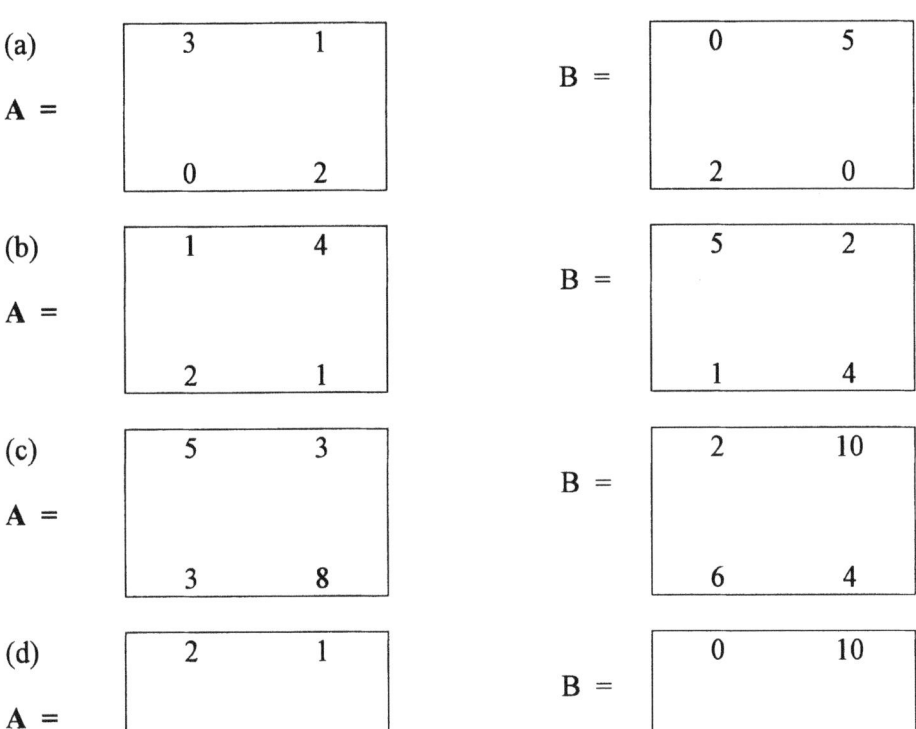

2. Consider an economy consisting of two kinds of goods: fuel and transportation. (Raw materials and labor are assumed in plentiful supply.) Fuel must be transported 100 miles from the mine to a processing plant so that each ton of fuel requires 100 T.-mi. of transportation, In turn, each ton-mile of transportation requires 4 lb. of fuel. Find the equilibrium price and intensity vectors for this economy. (1 T. = 2000 lb.)

3. Consider once again the economy of problem 2 above, but assume this time that, in addition to the inputs A, each ton of fuel requires wages of 100 lb. of fuel and 10 T.-mi. of transportation, while each ton-mile of transportation requires wages of ½ lb. of fuel and 10 lb.-mi. of transportation. How does this change prices and the expansion factor?

4. In Problem 3, change the wages to 200 lb. of fuel and 5 T.-mi. of transportation per ton of fuel; and 1 lb. of fuel and 5 lb.-mi. of transportation for each ton-mile of transportation. How does this affect equilibrium prices?

10. TWO-PERSON NON-ZERO-SUM GAMES

Up to now, we have almost exclusively studied games in which the two players' interests were directly opposed. Though that theory has many applications, it is clear that there are very many situations in which the two players of a game can both gain (or both lose) depending on the course of action. Unfortunately, there is no such satisfactory theory for these games as there is for the strictly competitive (i.e., zero-sum) games studied so far.

These *non-zero-sum games* can generally be represented by two $m \times n$ matrices, $A = (a_{ij})$, and $B = (b_{ij})$. In this case, a_{ij} and b_{ij} are the amounts won by P_1 and P_2, respectively assuming that they choose the ith strategy (row) and jth strategy (column), respectively. (in the zero-sum case, we would find that $B = -A$.) Thus P_1 wishes to maximize a_{ij}, while P_2 wishes to maximize b_{ij}. Such games are called *bimatrix* games.

We consider two possibilities: cooperative games, in which such things as preplay communications, binding contracts, and side payments are allowed, and non-cooperative games, in which these are forbidden (e.g. by antitrust laws).

In the non-cooperative case, the "obvious" to look for equilibrium pairs. These are defined much as for matrix games.

IV.10.1. Definition. Let (A, B) be a bimatrix game. The ith row strategy and jth column strategy are an *equilibrium pair* if, for every k,

$$a_{ij} \geq a_{kj}$$

and, for every l,

$$b_{ij} \geq b_{il}.$$

Thus, for an equilibrium pair, a_{ij} is the largest element in its column, while b_{ij} is the largest element in its row. (Recall that in this case P_2 wishes to maximize b_{ij} rather than to minimize it.)

The principal question, now, is whether bimatrix games will have equilibrium pairs. The answer is exactly the same as for matrix games: there need be no equilibrium pair of pure strategies, but there will always be an equilibrium pair of mixed strategies. (We omit the proof of this theorem as it requires some rather advanced mathematics.) Unfortunately the behavior of equilibrium pairs for these games can be quite singular, as the following examples show.

IV.10.2. Example. A company, C_1, has a factory that can be used to process either an expensive blend or an inexpensive blend of coffee. A second company, C_2, can produce either an expensive percolator or an inexpensive coffee urn. The expensive coffee can be brewed in either the percolator or the urn, but the cheap coffee can only be brewed in the percolator. If C_1 makes the expensive blend and C_2 makes the urn, then C_1 makes a large profit and C_2 a small profit. If C_1 makes the cheap blend and C_2 makes the percolator, then it is C_2 that makes the large profit and C_1 the small profit. If both make the expensive products, then sales are very low and not enough to cover expenses. If both make the cheap products, there are no sales.

The matrices for this game might reasonably be assumed to look like this:

$$A = \begin{bmatrix} 4 & -1 \\ -1 & 1 \end{bmatrix} \qquad B = \begin{bmatrix} 1 & -1 \\ -1 & 4 \end{bmatrix}$$

This game has two equilibrium pairs in pure strategies, corresponding to one company making its cheap product, while the other makes its expensive product. (There is a third equilibrium, using mixed strategies, but we may safely ignore it.) It may be seen, however, that the behavior of equilibrium pairs is not nearly as satisfactory as for matrix games. First of all, it may be seen that C_1 prefers one equilibrium pair, while C_2 prefers the other (i.e., the pairs are not equivalent as they do not give the same payoffs). For this game, disclosure as to one's intentions may actually be advantageous. If C_1 can just make the announcement, "We will produce the expensive coffee, and that's that!" and make it sound convincing enough, then C_2 might just be forced to follow along and produce the cheap urns. (Of course, C_2 might believe that C_1 was only bluffing, which might lead to disaster for both of them.) Note also that, in the absence of preplay communications (possibly forbidden by antitrust laws), C_1 might aim for one equilibrium while C_2 aims for the other, giving a non-equilibrium result. For this game, equilibrium pairs are not interchangeable.

Even in a game that has only one equilibrium pair, the results might be rather unsatisfactory.

IV.10.3. Example (The Price War). (This example is the work of M. Dresher and A. W. Tucker.) Two companies produce an identical product, which may be sold at either of two prices: high or low. If both sell at high price, then they both make a small profit; if at a low price, both break even. If, however, one sells at a high price, and the other at a low price, then the company selling at the lower price makes a large profit, whereas the other one gets no customers and takes a loss. The game may be represented by the two matrices

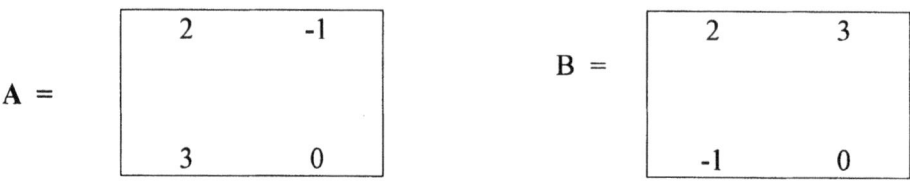

It may be seen that the second row of A dominates the first, while the second column of B dominates the first. The only equilibrium pair is (low price, low price). This gives a payoff of 0 units to each company. Yet both companies would do better if they were to use the *dominated* strategies (high price, high price) which gives them a profit of 2 units each. The difficulty is that neither can count on the other to maintain a high price, and price-fixing agreements are (by hypothesis) forbidden.

IV.10.4. Example (The Prisoner's Dilemma). A district attorney has two prisoners, whom he suspects of engineering a big robbery. Unfortunately he has no evidence, so that his only hope of obtaining a conviction is to have one of them confess.

Isolating the two prisoners, the D. A. talks to each one as follows: "I have enough information to send you both to jail on some minor charges, and I will do so if neither confesses. If you both confess, then you will both receive moderate sentences. But if only one of you confesses, then the one who confesses will be set free, whereas the other will receive a very long prison term. You had better confess before your partner does so."

We can reasonably assign utilities of 0 for being set free, -1 for the light sentence, -5 for the moderate sentence, and −10 for the long prison term. In this case, each prisoner has two strategies, "confess" and "do not confess." The game matrices are:

It may be seen that domination in this game is exactly as in Example IV.10.3, above. Once again, the only equilibrium point corresponds to second row, second column: both prisoners confess. Yet if they both confess, they do worse than if neither one confesses. Once again, we see that it may be better for both players to act "irrationally."

An obvious question is as to the computation of equilibrium pairs for bimatrix games.

In general, computation of these pairs is considerably more complicated than the computation of optimal strategies for matrix games. The reason is that this problem cannot be represented by a linear program; hence linear programming techniques cannot be used. Moreover, the method of fictitious play will not, in general, succeed here.

Of course, if an equilibrium pair exists in pure strategies, it is generally easy enough to find this. Otherwise, some rather complicated combinatorial mathematics must be used to find equilibrium pairs. For 2×2 games, however, the following is of interest:

If a 2×2 bimatrix game, (A, B) has an equilibrium pair in mixed strategies, then these equilibrium strategies can frequently be obtained by using the techniques of Section VI.4 above, with the understanding that P_1 chooses **x** as his optimal strategy in the matrix game B, while P_2 chooses **y** as his optimal strategy in matrix game A.

IV.10.5. Example. Consider the game

$$A = \begin{bmatrix} -1 & 5 \\ 2 & -5 \end{bmatrix} \quad B = \begin{bmatrix} 4 & 0 \\ -10 & 5 \end{bmatrix}$$

In this game, we can check directly that there is no equilibrium pair in pure strategies. In this case, we note that if P_1 were playing game B, his optimal strategy would be **x** = (15/19, 4/19). If P_2 were playing A, his optimal strategy would be **y** = (10/13, 3/13). Now, these turn out to form an equilibrium pair for (A, B). In fact, if P_1 plays **x**, then P_2's expected payoff will be 20/19, whatever he does. Similarly, if P_2 plays **y**, then P_1's expected payoff will be 5/13 whatever he does. Thus the pair (**x,y**) is in equilibrium.

IV.10.6. Example. Consider once again the game of Example IV.10.2. Even though we have already found two equilibrium pairs in pure strategies, we might look for an equilibrium pair in mixed strategies. In fact, the game is given by the matrices

$$A = \begin{bmatrix} 4 & -1 \\ -1 & 1 \end{bmatrix} \qquad B = \begin{bmatrix} 1 & -1 \\ -1 & 4 \end{bmatrix}$$

For game B, P_1's optimal strategy is $\mathbf{x} = (5/7, 2/7)$. In A, P_2's optimal strategy is $\mathbf{y} = (2/7, 5/7)$. It can be checked that these are in equilibrium for (A,B), giving payoffs of 3/7 to both players.

It should be noted that this equilibrium pair is far inferior to the two pure strategy pairs discussed above. The equilibrium property is not necessarily good in such cases.

COOPERATIVE GAMES

Up to this point, we have dealt with non-cooperative games, where such things as pre-play communication, contracts, side payments, and such are forbidden. As opposed to these, we may consider cooperative games. In the simplest of cooperative two-person games, the two players can obtain a certain amount of utility if they can come to some agreement; the main problem then is to decide how this utility is to be divided between the two players.

To take a trivial example, suppose some benefactor offers to give two players $200, to divide any way they wish, if only they can agree on this division. The "even split," $100 each, seems eminently fair, but there is now way to enforce this if one of the two players were to insist, say, on $150. In other words, every division of the money is in some way possible: if one of the two players needs $50 very badly, he may find himself forced to accept $50 rather than nothing at all.

The search in such games is for a "fair" solution -- an outcome that will adequately represent the players' bargaining *positions*, though not their bargaining *abilities*. It may be thought of as an arbitrated outcome: the outcome that a fair and impartial arbitrator might give.

The *arbitrated threat solution* is obtained in the following way: given the two matrices A and B, find the *threat differential*, which is merely the value of the game A-B. Let this value be v. This represents the relative advantage which P_1 is able to obtain over P_2. Then the arbitrated threat solution divides the maximal possible utility that can be obtained by the two players in such a way that P_1 obtains v more units than P_2. In other words, the solution divides the possible gains so as to maintain the relative advantage v.

Note, however, the following: it is possible that those strategies which maximize the sum of the payoffs may give too much to one of the two players. In such case, this player would be expected to make a *side payment* to the other, keeping the sum of the two payoffs fixed, but changing the division of the payoffs. The reason for this is clear: this is the price of obtaining cooperation.

IV.10.7. Example. Two companies, C_1 and C_2, produce equivalent automobiles. Neither can produce enough to saturate the market, but, if both produce as much as possible, they can oversaturate the market.

If one company produces its maximum while the other produces only a small amount, then the one that produces a lot makes a large profit, while the other makes a small profit. If both produce to capacity, there is a glut on the market, and they both lose money. If both produce small amounts, then there is a danger of a recession, which will cause losses to both companies. The situation is asymmetrical because C_1 has recently bought some very expensive and efficient machinery which enables it to produce at somewhat lower cost but must, on the other hand, be amortized. Hence C_1 suffers a large loss if thee is underproduction, but only a small loss if there is overproduction, while C_2's loss is large for overproduction and small for underproduction.

The game can be represented by the matrices

$$A = \begin{bmatrix} -1 & 4 \\ 1 & -4 \end{bmatrix} \qquad B = \begin{bmatrix} -4 & 1 \\ 4 & -1 \end{bmatrix}$$

where the first row and first column represent high production; the second row and second column, low production.

At first glance, the games seems to be quite symmetric, which suggests that both companies should produce a moderate amount, giving a total utility (to both) of 5 units; each would then receive 5/2 units of profit. On the other hand, if we form the "threat matrix"

$$A - B = \begin{bmatrix} 3 & 3 \\ -3 & -3 \end{bmatrix}$$

we see that the threat differential is $v = 3$. The maximal possible utility available to both companies is obtained by taking the matrix

$$A + B = \begin{bmatrix} -5 & 5 \\ 5 & -5 \end{bmatrix}$$

whose maximal entry is 5. There are 5 units of utility, to be divided between C_1 and C_2, in such a way as to maintain C_1's threat advantage of 3 units. This gives us the arbitrated threat solution $(4,1)$: C_1 should produce as much as it can, while C_2 keeps its production low.

IV.10.8. Example. Consider the cooperative bimatrix game

$$A = \begin{bmatrix} -1 & 2 \\ 1 & -4 \end{bmatrix} \qquad B = \begin{bmatrix} -4 & 1 \\ 4 & -1 \end{bmatrix}$$

For this game, the threat differential is given by the matrix game

$$A - B = \begin{bmatrix} 3 & 1 \\ -3 & -3 \end{bmatrix}$$

which has value $v = 1$ (note the saddle point at the upper right-hand entry).

We also find

$$A+B = \begin{bmatrix} -5 & 3 \\ 5 & -5 \end{bmatrix}$$

so that the two players can obtain a total of five units by using the second row and first column, respectively. Note, however, that this choice of strategies would give P_1 only 1 unit, while giving 4 units to P_2. This is not in keeping with the threat differential, which says that P_1 should receive 1 unit more than P_2. In such a case, we can imagine P_1 threatening to use row 1 unless he receives somewhat more. To avoid the threat, P_2 will sign an agreement, offering to make a side payment of 2 units to P_1 as an inducement to cooperation. With this side payment, P_1 will receive a total of 3 units, while P_2 gets 2 units – thus maintaining the difference $v = 1$.

PROBLEMS ON TWO-PERSON NON-ZERO-SUM GAMES

1. Find equilibrium pairs for the following bimatrix games:

(a)
$$A = \begin{bmatrix} 3 & 5 \\ 2 & 1 \end{bmatrix} \qquad B = \begin{bmatrix} 1 & 4 \\ 8 & 2 \end{bmatrix}$$

(b)
$$A = \begin{bmatrix} 6 & 1 \\ 1 & 6 \end{bmatrix} \qquad B = \begin{bmatrix} 2 & 1 \\ 1 & 2 \end{bmatrix}$$

(c)
$$A = \begin{bmatrix} 2 & 3 \\ 5 & 1 \end{bmatrix} \qquad B = \begin{bmatrix} 4 & 3 \\ 1 & 5 \end{bmatrix}$$

(d)

$A = \begin{bmatrix} 1 & 5 \\ 3 & 2 \end{bmatrix}$ $B = \begin{bmatrix} 2 & 3 \\ 4 & 1 \end{bmatrix}$

(e)

$A = \begin{bmatrix} 4 & -1 \\ -1 & 1 \end{bmatrix}$ $B = \begin{bmatrix} 1 & 6 \\ 2 & -3 \end{bmatrix}$

2. *The Inspector's Game.* (This example is due to M. Maschler.)

Player I, an *inspector*, tries to prevent player E, an *evader*, from carrying out some forbidden action. E can carry out the action in either of two time periods or not at all. I can inspect in either of those two time periods, but he is only allowed to inspect twice. If I inspects at the same time E tries to act, then E will certainly be caught. We assume a payoff (to I) of 1 if E does not act, 0 if E acts and is caught, and −1 if E acts and escapes detection. The payoff to E is 1 if he acts successfully (i.e., is not caught), 0 if he does not act, and −1 if he is caught.

    (a)   Show that this game can be represented by the matrices

$A = \begin{bmatrix} 0 & -1 \\ -1 & 1 \end{bmatrix}$ $B = \begin{bmatrix} -1 & 1 \\ 1 & 0 \end{bmatrix}$

    (b)   Show that $\mathbf{x}^* = (1/3, 2/3)$ and $\mathbf{y}^* = (2/3, 1/3)$ form an equilibrium pair. What are the expected payoffs in this case? (In fat, this is the only equilibrium pair.)

    (c)   Show that if I "announces" a strategy $(1/3+\varepsilon, 2/3-\varepsilon)$, e.g., $(0.34, 0.66)$, E's best strategy is $(0,1)$. Show that this gives a better payoff to I than the equilibrium pair $(\mathbf{x}^*, \mathbf{y}^*)$.

3. In each of the bimatrix games of problem 1, above, find the threat differential and the threat solution. State whether side payments are necessary.

11. EVOLUTIONARY STABLE SYSTEMS

We treat in this section an interesting application of game theory to biology. This is the *evolutionary stable system*.

As an example, let us suppose that there is a certain species of animal, whose males can be classified as either pugnacious or peaceful. We will call these, respectively, *hawks* and *doves*. Note, however, that these are not two different species of birds: they are two *types* of the same species of animal (not necessarily birds), and they are generally indistinguishable except for their behavior in competitive situations.

Suppose, then, that a competitive situation arises, in which two of these males meet, possibly to compete for some prize (food, shelter, or perhaps a female of the species). If, now, one of them is a hawk, while the other is a dove, then the dove will yield to the hawk without a fight. Assign, then, a payoff of +25 to the hawk, and 0 to the dove.

If two hawks meet, then they will fight until one of them is unable to continue – possibly dead, or at least near death. We might assign a payoff of +15 to the winner, and –35 to the loser. Since either has probability 0.5 of winning, each has expected payoff –10.

Finally, if two doves meet, they will probably try to scare each other off, but since they are doves they will eventually come to some distribution of the goods. This might give them an expected payoff of +10 each.

This situation gives rise to the bimatrix game

$$A = \begin{array}{c|cc} & D & H \\ \hline D & 10 & 0 \\ H & 25 & -10 \end{array} \qquad B = \begin{array}{c|cc} & D & H \\ \hline D & 10 & 25 \\ H & 0 & -10 \end{array}$$

in which the first row and column correspond to a dove, and the second row or column correspond to a hawk. Note that the game is symmetric in the sense that $B = A^t$.

We see that this bimatrix game has two equilibria in pure strategies: (D,H) and (H,D): the hawk gets everything without a fight, and the dove gets nothing but avoids a dangerous fight. There is also an equilibrium pair of mixed strategies, in which both players use $\mathbf{x} = (0.4, 0.6)$, yielding an expected payoff of 4 units to each player.

What we need to understand now is the meaning of these equilibrium pairs. If the two players in this game were humans – a pair of teen-agers, perhaps, playing "chicken" – it would be easy to understand the (H,D) and (D,H) equilibria. In each of these, one of the two players fights, the other runs away. (Compare the situation with that of Example IV.10.2, above, whose equilibria are very similar to these.)

In the present case, however, it is difficult to see just how such an equilibrium can be implemented. The equilibrium (H,D) says that P_1 should fight while P_2 runs away. But – since these animals are not assumed rational – this can only happen if P_1 is a hawk while P_2 is a dove. Since all of the given animals are assumed equally likely to take part in the given type of competitive situation, it is not possible to arrange things so that all these matches will pit a hawk versus a dove. Thus, the (H,D) equilibrium pair is impossible to implement, as is the (D,H) pair.

Consider, on the other hand, the mixed strategy equilibrium, $\mathbf{x} = \mathbf{y} = (0.4, 0.6)$. If, in the general population, 40% of the animals are doves, while the other 60% are hawks, then it is clear that, in a random matching of these males, each of the two will have 0.4 probability of being a dove and 0.6 of being a hawk. Thus, a statistically inclined observer might well think that in every such match, the animals were actually rational players using the given mixed strategy. Note that this is possible only because the equilibrium is a *symmetric* pair, i.e. $\mathbf{x} = \mathbf{y}$.

The next question is as to why nature should choose this fraction of doves and hawks in the general population. The answer is that – according to a well-known theory – species *evolve* in such a way as to optimize their probability of survival.

In fact, suppose that, in this particular species of animal, 30% of the males were doves, and 70% were hawks. Treating this as a mixed strategy $\mathbf{x} = (0.3, 0.7)$, we find that

$$\mathbf{x}B = (3, 0.5)$$

and this tells us that if, in this case, P_1 is taken at random from the general population, then P_2 will gain more if he is a dove than if he is a hawk. According to the theory of evolution, this would make doves more likely to survive, and – since this trait is presumably inheritable – the proportion of doves in the population would increase.

Contrariwise, if the population showed 50% of each type of male, then setting $\mathbf{x} = (0.5, 0.5)$, we find

$$\mathbf{x}B = (5, 7.5)$$

which means that hawks would survive and breed more than doves, and so the fraction of doves would decrease. More generally, we find that, starting from any population mixture other than (0.4, 0.6), there will be a tendency to move towards this particular mixture. Thus, the equilibrium represents a stable situation: the system will attempt to return there, correcting any disruptions.

The reader may feel that, in the strict sense of the word, the equilibrium does not maximize the species' survivability. In fact, the expected payoff in case of the equilibrium is only 4 units to each player. A species containing only doves would give us the (D,D) payoff or 10 units to each. Would this (a totally peaceful society) not be better?

The answer is that the species needs some hawks, not to make life better for the doves, but, rather, as protection against other hawks. In fact, suppose that a tribe of animals of this species consisted entirely of doves. In this case, it would be subject to invasion by a "foreign" tribe which contained a good number of hawks. These hawks would serve as warriors for the invading tribe, which would have no difficulty then in driving the formerly established tribe away. The original tribe, then, needs its own hawks so as to avoid such invasion.

Would it then be better to have a tribe consisting only of hawks? The answer is, once again, no: too many hawks would cause too much internecine fighting within the tribe. In such an atmosphere (in which the overwhelming majority of males were hawks), a small number of doves would actually do better by avoiding fights. This is why the equilibrium (0.4, 0.6), and no other mixture, is considered stable.

In this last example, there was only one symmetric equilibrium. Consider, however, the following:

IV.11.2. Example. Two automobiles approach each other along an otherwise empty road. Both are traveling along the center of the road, but the road is wide enough for the automobiles to pass. Each car has the alternative of moving either to the left or to the right of the road. If both move to the right, or both to the left, they will pass without a problem. On the other hand, if one moves right while the other moves left, there will be a serious accident.

The situation can be represented by the bimatrix game

$$A = \begin{array}{c|cc} & L & R \\ \hline L & 10 & -10 \\ R & -10 & 10 \end{array} \qquad B = \begin{array}{c|cc} & L & R \\ \hline L & 10 & -10 \\ R & -10 & 10 \end{array}$$

and it is easy to see that there are two equilibrium pairs in pure strategies, namely (L,L) and (R,R). There is also an equilibrium pair in mixed strategies, in which both players use $\mathbf{x} = \mathbf{y} = (0.5, 0.5)$.

It is interesting to note that both (L,L) and (R,R) give the same payoff of +10 to all players. In fact, certain countries have adopted the L strategy, some have adopted R, and there is no reason why one should be preferred over the other. Both of these equilibria are stable, in the following sense:

Suppose in some society (e.g. Britain or Japan), people have agreed to drive on the left side of the road. Suppose then that a number of people from a right-side driving society (say, the United States or France) arrive in this society and decide to continue driving on the right. So long as this number of right-side drivers is small relative to the total number of drivers, we find that, on most occasions when these right-side drivers meet another automobile, they will have accidents and be removed from the road. Eventually, they will realize what the situation is and decide to change from right-side to left-side driving. Of course, the number of left-side drivers involved in accidents will be just as large, but since there were many more left-side drivers to begin with, the number of right-side drivers, *as a fraction of the total number of drivers*, will decrease until eventually only a few are left. Thus the society will move back toward the equilibrium in which all motorists drive on the left side of the road. The conclusion is that, if there is small perturbation from the original equilibrium, the society will tend to return to this equilibrium. It is in this sense that the equilibrium is stable.

In a similar way, the equilibrium (R,R) is also stable.

Consider, on the other hand, the equilibrium with $\mathbf{x} = \mathbf{y} = (0.5, 0.5)$. The expected payoff here is 0 units for all drivers. In this case, half the motorists will drive on the left, and half on the right. This is an equilibrium in the sense that, since an oncoming driver is as likely to move left as to move right, no driver has any incentive to change his own tendency.

Suppose, however, that some small (but not negligible) number of motorists arrives in this society, all coming from a right-side driving society. This will, let us say, change the proportion of motorists to $\mathbf{z} = (0.48, 0.52)$. Now, at this point, some drivers will notice that the chance of an accident is slightly larger for left-side than for right-side drivers (0.52 to 0.48 probability). This will cause those who were previously driving on the left to switch strategy, so that after a while we might find that the fraction of right-side drivers has increased to something like 60%. But this will be even more reason to switch to the right side, and eventually the society will reach the (R,R) equilibrium. Thus, the mixed strategy equilibrium is unstable, in the sense that the society does not have a tendency to return to it after small perturbations.

In general, we can say that an equilibrium is *stable* if there is a tendency to return to it after *small* perturbations. In other words, if the equilibrium mixture changes slightly – either because of invasion by some foreign tribe of animals of the same species, or because of mutations -- the environment will be less hospitable to the invaders (or mutants) than to the original society, so that the invaders eventually have the choice of conforming to the original equilibrium, or disappearing. This can actually come about in two different ways:
(a) the original environment is inhospitable to the invaders, or
(b) the new environment (changed by the presence of the invaders) is inhospitable to the invaders.

In Example IV.11.1 above, we find that case (b) holds: the equilibrium (0.4, 0.6) environment is equally hospitable to all, but the change to a different mixture (by an invasion of a tribe containing either more than 60% hawks, or more than 40% doves) will prove inhospitable to the invaders.

In Example IV.11.2, it is case (a) that holds: the right-side driving environment will be inhospitable to left-side drivers.

This leads us to the following definition of an *evolutionary stable system*:

IV.11.3. Definition. A bimatrix game (A,B) is *symmetric* if $B = A^t$. For such a game, an equilibrium pair of strategies $(\mathbf{x}, \mathbf{y})$ is *symmetric* if $\mathbf{x} = \mathbf{y}$.

IV.11.4. Definition. Let (A, A^t) be a symmetric bimatrix game. An evolutionary stable system (E. S. S.) is a (mixed) strategy $\mathbf{x}$ such that:

(a) $(\mathbf{x}, \mathbf{x})$ is a symmetric equilibrium pair

(b) if $\mathbf{z}$ is any other mixed strategy, then either

(i) $\mathbf{x}A\mathbf{x}^t > \mathbf{z}A\mathbf{x}^t$

or

(ii) $\mathbf{x}A\mathbf{z}^t > \mathbf{z}A\mathbf{z}^t$

The question now arises as to the existence of E. S. S.'s. In fact, it turns out that, with mixed strategies, any symmetric bimatrix game has at least one symmetric equilibrium pair. However, not all symmetric equilibrium pairs are E. S. S.'s, and in general not all symmetric games have E. S. S.'s.

In dealing with pure strategies, the following result is interesting:

IV.11.5. Theorem. Let (A, A^t) be a symmetric bimatrix game. If there is some i such that $a_{ii} > a_{ji}$ for all $j \neq i$, then the ith pure strategy is an E.S.S.

IV.11.6. Example. Returning to Example IV.11.2., we note that $a_{11} = 10$ while $a_{21} = -10$. Thus the first pure strategy (L) is an E. S. S. The same is true for the second pure strategy.

If, however, we consider the mixed strategy $\mathbf{x} = (0.5, 0.5)$, we note that

$$A\mathbf{x}^t = (0, 0)^t$$

and so

$$\mathbf{x}A\mathbf{x}^t = \mathbf{z}A\mathbf{x}^t = 0 \quad \text{for all } \mathbf{z}.$$

Now, if we let $\mathbf{z} = (1, 0)$, we find that

$$\mathbf{x}A\mathbf{z}^t = 0 < \mathbf{z}A\mathbf{z}^t = 1$$

so that condition (b) does not hold, and this equilibrium, though symmetric, is not stable.

IV.11.7. Example. To see that not all symmetric games need have E. S. S.'s, consider the game

$$A = \begin{bmatrix} 2 & 1 \\ 2 & 1 \end{bmatrix} \qquad B = \begin{bmatrix} 2 & 2 \\ 1 & 1 \end{bmatrix}$$

Since the two rows of A are identical, it follows that P_1's payoff depends only on P_2's choice. Thus, any response of P_1 will be optimal against whatever P_2 does. By symmetry, P_2's payoff depends only on P_1's strategy; and it follows that for any mixed strategy $\mathbf{x}$, the pair $(\mathbf{x},\mathbf{x})$ is a symmetric equilibrium. However, there is no E. S. S.

To see that there is no E. S. S., we note that, since P_1's payoff depends only on P_2's choice, we will have, for any $\mathbf{x}$ and $\mathbf{z}$,

$$xAx^t = zAx^t$$

and

$$xAz^t = zAz^t$$

so that **x** cannot be an E.S.S.

PROBLEMS ON EVOLUTIONARY STABLE SYSTEMS

Find symmetric equilibria for each of the following games. Which of these are E. S. S.'s?

(a)
$$A = \begin{bmatrix} 3 & 5 \\ 2 & 4 \end{bmatrix} \qquad B = \begin{bmatrix} 3 & 2 \\ 5 & 1 \end{bmatrix}$$

(b)
$$A = \begin{bmatrix} 4 & 2 \\ 5 & 1 \end{bmatrix} \qquad B = \begin{bmatrix} 4 & 5 \\ 2 & 1 \end{bmatrix}$$

(c)
$$A = \begin{bmatrix} 8 & 0 \\ 11 & 2 \end{bmatrix} \qquad B = \begin{bmatrix} 8 & 11 \\ 0 & 2 \end{bmatrix}$$

(d)
$$A = \begin{bmatrix} 4 & 7 \\ 5 & 1 \end{bmatrix} \qquad B = \begin{bmatrix} 4 & 5 \\ 7 & 1 \end{bmatrix}$$

(e)
$$A = \begin{bmatrix} 4 & 3 \\ 2 & 5 \end{bmatrix} \qquad B = \begin{bmatrix} 4 & 2 \\ 3 & 5 \end{bmatrix}$$

V. COOPERATIVE GAMES

1. N-PERSON GAMES

For cooperative games with more than two players, a new complication enters into games, namely that of coalition formation. Specifically, we note that, for two-person zero-sum games, there is no point in cooperation. For two-person general-sum games, if cooperation is allowed, the two players have a choice of cooperating or not. For n-person games (where $n \geq 3$), we find that, in most cases, players have a choice of joining one or another coalition. It is this process of coalition-formation, and the consequent bargaining possibilities, that we will now study.

We shall label the players $1, 2, \ldots, n$. Then $N = \{1, 2, \ldots, n\}$ is the *set of all players* in a game. All the non-empty subsets $S \subset N$, including N itself, and all the single-element sets, will be called *coalitions*. We assume that each coalition has certain strategies (joint strategies of its several members) which it can use. We also assume that it would know how best to use these strategies, so as to maximize the amount of utility received by all of its members. Thus, we will dispense with the normal form of the game and study, instead, the *characteristic function* form. This is a function v which tells us the maximum that each coalition can obtain.

V.1.1. Definition. The characteristic function of an n-person game is a real-valued function, v, whose domain is the collection of all subsets of $N = \{1, 2, \ldots, n\}$, satisfying

5.1.1 $$v(\emptyset) = 0$$

5.1.2 $$v(S \cup T) \geq v(S) + v(T) \quad \text{if } S \cap T = \emptyset$$

The quantity $v(S)$ is the amount of utility that S can obtain. We assume that S can then divide this utility among its members in any way that they see fit. Condition (5.1.1) says simply that the empty set can obtain nothing for its (non-existent) members. Condition (5.1.2) is known as *super-additivity* and represents the fact that coalitions S and T, acting together, can obtain at least as much utility for their members as when acting separately.

V.1.2. Example. Consider the following three-person game. Each of three players chooses either heads (H) or tails (T). If two players choose the same, but the remaining player chooses differently, then this last player must give 1 unit to each of the other two players. In any other case, there is no payoff.

For this game, it is clear that
$$v(\varnothing) = 0$$

Since the empty set can obtain nothing, and
$$v(\{1,2,3\}) = 0$$

since the game is zero-sum: the three players together can gain nothing since all payments must take place among them.

Suppose, next, that the coalition {2,3} forms. It will be opposed by player 1, giving what is essentially a two-person game between the two coalitions, {1} and {2,3}. The coalition {1} has the two strategies H and T, while {2,3} has four strategies, HH, HT, TH, and TT. This gives us a 2×4 matrix game

	HH	HT	TH	TT
H	0	1	1	-2
T	-2	1	1	0

For this matrix, the second and third columns are dominated, and it is easy to see that the value of the game is –1. This is what the maximizing party (player 1) expects to gain. The minimizer in this game (coalition {2,3}) then expects to gain 1 unit. Thus

$$v(\{1\}) = -1$$

$$v(\{2,3\}) = 1$$

By the symmetry of this game, we conclude that the characteristic function is

$$v(S) = \begin{cases} 0 & \text{if } |S| = 0 \text{ or } 3 \\ -1 & \text{if } |S| = 1 \\ +1 & \text{if } |S| = 2 \end{cases}$$

where $|S|$ is the number of players in coalition S.

This example shows how the characteristic function of a game may be obtained from its normal form. Very often, however, the game is given directly in characteristic function form.

V.1.3. Example. In a three-person game, player 1 (the *seller*) owns a horse which is worthless to him. Player 2 (a *farmer*) and player 3 (a *butcher*) would both like to buy the horse. The farmer feels that the horse is worth $100, while the butcher feels that the horse is worth $50.

In this game, the general idea is that, by transferring ownership of the horse from the seller to one of the two buyers, utility is created. Thus, coalition {1,2} can create 100 units (dollars) of utility, which the two players can divide in any way they choose. For example, if they decide on a price of $30, then player 1 gains 30 units (he has exchanged a worthless horse for $30), while player 2 gains 70 units (he has obtained a $100 horse for $30.) Similarly, coalition {1,3} can gain 50 units of utility. Single-player coalitions obtain no utility; nor can {2,3} obtain any utility.

It is also possible that coalition {1,2,3} form: if so, the best they can do is to give the horse to player 2, and then a total of 100 units will be obtained. We thus have the characteristic function

$$v(\{1\}) = v(\{2\}) = v(\{3\}) = v(\{2,3\}) = 0$$
$$v(\{1,3\}) = 50$$
$$v(\{1,2\}) = v(\{1,2,3\}) = 100.$$

Let us suppose that a game, v, is played. What payoffs can the players expect from this game? This is not an easy question to answer – there is no generally accepted theory to determine a unique *solution* to such a game. We may assume, however, that no player will accept less than he could obtain for himself (acting alone). We will also assume that the players are able to come to an agreement so that they can share all the available utility. This leads us to the concept of an *imputation*.

V.1.4. Definition. Let v be the characteristic function of an n-person game. An imputation is a vector $\mathbf{x} = (x_1, x_2, \ldots, x_n)$ satisfying

5.1.3
$$\sum_{i=1}^{n} x_i = v(N)$$

5.1.4
$$x_i \geq v(\{i\}) \quad \text{for all } i.$$

Conditions (5.1.3) and (5.1.4) are known respectively as *group rationality* and *individual rationality*. In essence, an imputation is one of the possible distributions of the profits (payoffs) that might be obtained if game v is played.

In the game of Example V.1.2, the imputations are vectors $\mathbf{x} = (x_1, x_2, x_3)$ satisfying $x_i \geq -1$ and $x_1 + x_2 + x_3 = 0$. In that of Example V.1.3, they are vectors $\mathbf{y}$ satisfying $y_i \geq 0$ and $y_1 + y_2 + y_3 = 100$.

Unfortunately, most games have an infinity of imputations, and it is not clear how to distinguish one, or at least a small subset, of these imputations, as being most "reasonable." Several "solution concepts" have been suggested for n-person games; we shall consider only two, namely the *core* and the *power index*.

2. THE CORE

V.2.5. Definition. The *core* of an n-person game v is the set of all imputations $\mathbf{x} = (x_1, x_2, \ldots, x_n)$ such that, for every coalition S,

5.2.5
$$\sum_{i \in S} x_i \geq v(S)$$

Note what this means: if an imputation belongs to the core, then every (potential or actual) coalition is obtaining as much as it possibly could obtain for its members if it were to form. Thus a core imputation exhibits a great deal of stability. If, on the other hand, an imputation is not in the core, then there is at least one coalition S whose members are not getting as much as they could; there will then be strong pressure for change on the part of the members of S.

Let us look, then, for the cores of the games just described. In Example V.2.2, it turns out that the core is empty. For, suppose that the imputation $\mathbf{x}$ were in the core. Then, since $v(\{1,2\}) = 1$, we must have

$$x_1 + x_2 \geq 1$$

But, because $\mathbf{x}$ is an imputation,

$$x_1 + x_2 + x_3 = 0.$$

But this means that $x_3 \leq -1$. But by a similar argument, we find that $x_1 \leq -1$ and $x_2 \leq -1$. But this contradicts the hypothesis that $x_1 + x_2 + x_3 = 0$. In other words, $\mathbf{x}$ cannot be an imputation. We conclude that there are no imputations in the core. This may be interpreted as meaning that the game is unstable; it may also be interpreted as meaning that the core is not a totally satisfactory solution concept.

Consider next Example V.1.3. If **x** is an imputation, then its components must all be non-negative, and

$$x_1 + x_2 + x_3 = 100.$$

If it is in the core, we must also have

$$x_1 + x_2 \geq 100$$

which means that $x_3 \leq 0$. But since all components are non-negative, we must have $x_3 = 0$. Then $x_1 + x_2 = 100$. But, again by the core conditions,

$$x_1 + x_3 \geq 50.$$

Since $x_3 = 0$, it follows that $x_1 \geq 50$. On the other hand, x_2 must also be non-negative, and so $x_1 \leq 100$. We conclude that an imputation in the core must be of the form

$$\mathbf{x} = (x_1, 100-x_1, 0) \quad \text{with} \quad 50 \leq x_1 \leq 100.$$

We can give an interesting interpretation to this. Essentially, the farmer will buy the horse for a price somewhere between $50 and $100. The price cannot be greater than $100 since the horse is only worth that much. On the other hand, the price must be at least $50 since, if the farmer offers less than $50, then the butcher can make a competing offer. Note that the butcher will get nothing from the game; nevertheless, his presence is important in "bidding up" the price. In effect, the butcher gives the seller a choice of coalition partner; the farmer has no such alternative coalition to join and this is reflected in the seller's better bargaining position and consequent greater gain in the game.

V.2.6. Example. Let R and L be two disjoint sets of players, such that $R \cup L = N$. Each member of R has one right-hand glove, while each member of L has one left-hand glove. Assume all the gloves are identical except for the left-right difference. Assume also that a pair of gloves is worth 1 unit (which can be whatever the market price of a pair of gloves might be), whereas single gloves have no market value. It is then easy to see that, for any coalition S,

$$v(S) = \min\{|S \cap R|, |S \cap L|\}$$

(where, once again, $|S \cap R|$ means the number of elements of set $S \cap R$, etc.) In particular, we note that $v(\{i\}) = 0$ for any single player; it is only by making coalitions (including both right-hand and left-hand gloves) that any utility can be obtained. We also have

$$v(N) = \min\{|R|, |L|\}.$$

Now, for an imputation **x** to be in the core, the condition (6.12.5) tells us that each right-hand glove owner and each left-hand glove owner must, together, receive at least one unit.

Assume first that the number of right-hand gloves is the same as the number of left-hand gloves, i.e. $|R| = |L| = m$. Then there are m units to be divided, and it is easy to see that, for any p with $0 \leq p < 1$, the vector **x** such that

$$x_i = \begin{cases} p & \text{if } i \in R \\ 1-p & \text{if } i \in L \end{cases}$$

is in the core. There are no other imputations in the core.

It is again easy to give an economic interpretation to this result: in the core, all right-hand gloves are paid at the price p, and the left-hand gloves of course receive $1-p$. But note that p is anything between 0 and 1: the price $p = \frac{1}{2}$ may seem more natural or fair, but there is no way it can be enforced.

Suppose, next, that there are fewer right-hand gloves than left-hand gloves: $|R| < |L|$. In this case, the core will consist of only one imputation, namely **x** such that

$$x_i = \begin{cases} 1 & \text{if } i \in R \\ 0 & \text{if } i \in L \end{cases}$$

We see that, in this case, the surplus of left-handed gloves makes them worthless. Note this will happen even if the difference in the number of gloves is minimal; i.e. even if we had 1000 right-handers and 1001 left-handers. The reason for this is that, in any attempt to form pairs of gloves (which is the only way that utility can be obtained) there will always be some left-handers unpaired, and their owners will presumably offer them at extremely low prices rather than remain unpaired.

In a similar way, if $|R| > |L|$, we find that it is the left-hand gloves that receive all the utility while the right-hand gloves become worthless.

The core is a very useful solution concept: in many cases it corresponds to the results of classical economic analysis. Unfortunately, it is very frequently empty, as can be seen from Example V.2.2 above.

PROBLEMS ON *n*-PERSON GAMES

1. in an n-person game v, an imputation **x** *dominates* another imputation, **y**, if there exists a coalition S such that

    (a) $\quad\quad\quad\quad\quad\quad x_i > y_i$ for every $i \in S$

    (b) $\quad\quad\quad\quad\quad\quad \sum_{i \in S} x_i \leq v(S)$

    (The notation for this is $\mathbf{x} \succ \mathbf{y}$. It should be remembered that S, being a coalition, is non-empty.)
    Prove that an imputation is in the core if and only if it is undominated.

2. For a 3-person game, let $v(\{i\}) = 0$ for all i, let $v(\{1,2\}) = v(\{1,3\}) = 60$, $v(\{2,3\}) = 75$, and $v(N) = 100$. Find a point in the core.

3. For a 3-person game, let $v(\{i\}) = 0$ for all i, let $v(\{1,2\}) = v(\{1,3\}) = 70$, $v(\{2,3\}) = 75$, and $v(N) = 100$. Prove that the core of this game is empty. [Hint: follow the proof of emptiness of the core in Example VI.12.2.]

4. More generally, for a 3-person game, let $v(\{i\}) = 0$ for all i, and $v(N) = 100$. Show that the game will have a non-empty core if and only if

    $$v(\{1,2\}) + v(\{1,3\}) + v(\{2,3\}) \leq 200.$$

5. For an *n*-person game, suppose that $v(S)$ depends only on $s = |S|$, the number of players in the set S. Show that the core is non-empty if and only if, for every s $(1 \leq s \leq n)$,

$$v(S) \leq s\, v(N)/n.$$

3. THE SHAPLEY VALUE

The second solution concept that we will study is the *power index* of L. S. Shapley, also known as the *Shapley Value*.

This value should be thought of as a function, φ, which assigns to each n-person game, v, an n-vector $\varphi[v]$. This vector represents a "reasonable" *a priori* expectation for the payoff of the game; it assumes that the game is adequately represented by the characteristic function alone. It also disregards such intangibles as personal affinities among the players and differences in bargaining ability.

The Shapley value is developed axiomatically, as follows.

Let T be any coalition. Define a game, u_T, by

5.3.1
$$u_T(S) = \begin{cases} 0 & \text{if } T \not\subset S \\ 1 & \text{if } T \subset S \end{cases}$$

For this game, known as the *unanimity-on-T* game, it is clear that the only question of importance is whether the members of the coalition T can all agree: these, acting together, can obtain 1 unit. The other players are *dummies*: they contribute nothing to any coalition. A "reasonable" expectation for this game would be $\varphi[u_T]$, given by

5.3.2
$$\varphi_i[u_T] = \begin{cases} 0 & \text{if } i \notin T \\ 1/t & \text{if } i \in T \end{cases}$$

where t is the number of elements in T. I.e., the unit should be divided in equal parts among the members of coalition T. Other players receive nothing.

Suppose, next, that v and w are n-person games in characteristic function form. If α and β are positive numbers, then the function $\alpha v + \beta w$ is also the characteristic function of an n-person game, and it is reasonable that

5.3.3
$$\varphi[\alpha v + \beta w] = \alpha \varphi[v] + \beta \varphi[w]$$

Condition (5.3.3) states two things: first, that if we multiply the stakes in a game uniformly by the constant α, the each player's expected winnings will also be multiplied by α; second, that the expected payoffs from playing the two games, αv and βw, simultaneously, should be the sum of the expected payoffs from playing the two games individually.

These two conditions, (5.3.2) and (5.3.3) are reasonable enough, on the surface, that we may treat them as axioms. Now it may be shown that each n-person game can be expressed as the sum and difference of games such as u_T (defined by (5.3.1)), multiplied by constants. For these games u_T, (5.3.2) defines a value $\varphi[u_T]$; condition (5.3.3) is then used to extend this definition to all n-person games.

Consider, alternatively, the following probabilistic model: take all the $n!$ possible orderings (permutations) of the n players in a game, and assign probability $1/n!$ to each. For each ordering, form a payoff vector in the following way: if player i is preceded by the members of coalition S (and no others) then i receives the amount

$$v(S \cup \{i\}) - v(S)$$

Then the expected value of this payoff is exactly equal to the Shapley value. In fact, it is trivial to verify (5.3.2); (5.3.3) follows from Theorem III.10.1.

(Note that we do *not* suggest that this probabilistic model is a realistic model of the way in which payoffs would actually be made; we introduce it only as an aid in computing the Shapley value.)

Using the probabilistic model, we can obtain a closed formula for the power index. In fact, we will compute the probability that, in a random ordering, player i will be preceded by the members of coalition S and no others. If S has s elements, then the members of S can be ordered in $s!$ different ways, and the members of N-S-$\{i\}$ (those who must follow i in the permutation) can be ordered in $(n-s-1)!$ different ways. There will then be $s!(n-s-1)!$ different permutations in which i is preceded by the members of S and followed by all others. The probability of such an ordering, then, is

5.3.4 $$\gamma_{n,s} = s!(n-s-1)!/n!$$

Thus, the Shapley value is given by the formula

5.3.5 $$\varphi_i[v] = \sum_{S \subset N} \gamma_{n,s} [v(S \cup \{i\}) - v(S)]$$

In this last formula, the summation is over all subsets S which do not contain player i. The value is *always* an imputation.

V.3.1. Example. Consider the "horse" game of Example V.1.4. Since $n = 3$, we first calculate

$$\gamma_{3,0} = \gamma_{3,3} = 1/3 \qquad \gamma_{3,1} = \gamma_{3,2} = 1/6$$

Now, for each i, there will be 4 summands in expression (6.13.4). We calculate

$$\varphi_1 = (1/3)\, 0 + (1/6)\, 100 + (1/6)\, 50 + (1/3)\, 50 = 58.33$$
$$\varphi_2 = (1/3)\, 0 + (1/6)\, 100 + (1/6)\, 0 + (1/3)\, 50 = 33.33$$
$$\varphi_3 = (1/3)\, 0 + (1/6)\, 50 + (1/6)\, 0 + (1/3)\, 0 = 8.33$$

and this gives us the value $\varphi = (58.33, 33.33, 8.33)$. Note that the value is not in the core, since player 3 (the butcher) gets something. This can be due to several things: for example, 2 and 3 might form a coalition (a buyer's syndicate) to drive the price down, and if so, 2 might make a side payment of a small amount to 3. Alternatively, 1 might have made a prior commitment to sell the horse to the butcher; when the farmer appears and offers a better price, 1 might find himself obliged to pay 3 so as to get avoid a law suit.

Formula (5.3.5) is complicated and, for large games, it can be quite difficult to apply. For $n = 3$ or 4, it is usually easier to take all the permutations of the players; since $3! = 6$ and $4! = 24$, the games are still small enough to do this. For larger games, it is sometimes possible to take advantage of some special feature in the structure of the game to simplify the calculations.

let us consider the game of Example V.2.3, with $|R| = 1$ and $|L| = 2$. In our model, the right-hand glove owner receives 1 unit if he is preceded in the ordering by at least one of the two left-hand glove owners. There are six possible orderings:

$$\begin{array}{ccc} 1,2,3 & 2,1,3 & 3,1,2 \\ 1,3,2 & 2,3,1 & 3,2,1 \end{array}$$

If player 1 is the owner of the fight-hand glove, we see that he obtains 1 unit in four of the six orderings, and 0 in the two others. Thus

$$\varphi_1 = 4/6 = 2/3.$$

Look next at player 2. He receives 1 unit only if he is preceded in the ordering by player 1 but not by player 3. There is only one such ordering, and so

$$\varphi_2 = 1/6$$

Similarly,

$$\varphi_3 = 1/6$$

In the general case, suppose there are r right-hand glove owners and l left-hand glove owners. A member of R receives 1 unit if he is preceded by more elements of L than elements of R, and 0 otherwise. Through some mathematics more complicated than can be considered here, it can be shown that, for $r \geq l$, the value to a player $i \in R$ is

$$\varphi_i = \tfrac{1}{2} - (r-l)/2r \sum_{k=0}^{l} r!\, l!\, /\, (r+k)!\, (l-k)!$$

while, for a player $j \in L$, it is

$$\varphi_j = \tfrac{1}{2} - (r-l)/2l \sum_{k=1}^{l} r!\, l!\, /\, (r+k)!\, (l-k)!$$

For small values of r and l, these values can be calculated directly, and are given in Table V.3.1. The entries represent the value to a member of R for each value of r and l.

TABLE V.3.1

r	l								
	0	1	2	3	4	5	6	7	8
1	0	0.500	0.667	0.750	0.800	0.833	0.857	0.875	0.889
2	0	0.167	0.500	0.650	0.733	0.786	0.822	0.847	0.867
3	0	0.083	0.233	0.500	0.638	0.720	0.774	0.811	0.867
4	0	0.050	0.133	0.272	0.500	0.629	0.710	0.764	0.802
5	0	0.033	0.086	0.168	0.297	0.500	0.622	0.701	0.755
6	0	0.024	0.060	0.113	0.194	0.315	0.500	0.616	0.693
7	0	0.018	0.044	0.081	0.135	0.214	0.330	0.500	0.610
8	0	0.014	0.033	0.061	0.099	0.153	0.230	0.341	0.500

As may be seen, the "short" end of the market has a definite competitive advantage. With three right-hand and five left-hand gloves, for example, the members of R receive 0.72 units each, or a total of 2.16. The members of L receive the remaining 0.84 units, or an average of 0.168 each. In a strictly competitive situation, however, as represented by the core, the members of L would receive nothing. The reason that the Shapley value gives L something is that it allows for the possibilities (a) that the left-hand glove owners might form a coalition among themselves to take some of the gloves off the market and thus raise their price and (b) that some right-hand glove owner might be unaware of the actual situation and thus sell his glove cheaply.

V.3.2. Example. A landlord owns a large piece of ground, which can be worked by peasants; there are $n-1$ peasants available and willing to work. It is assumed that if k peasants work the field, the production is $\sqrt{k}$ units.

We represent this as an n-person game in which player 1 is the landlord, and players 2, 3, ..., n are the peasants. We assume the landlord does not work. The peasants alone can produce nothing since they require land. Thus the characteristic function of the game is

$$v(S) = \begin{cases} 0 & \text{if } 1 \notin S \\ \sqrt{(s-1)} & \text{if } 1 \in S \end{cases}$$

where s is the number of elements in S.

We may compute the value to the first player, using the probabilistic model, in the following way: since the game treats all the other players equally, we need not concern ourselves with *which* players precede him in a given ordering, but only in *how many* players precede him. If he is preceded by k players, his marginal contribution is $\sqrt{k}$ units. For any k, $0 \leq k \leq n-1$, the probability that he is preceded by exactly k players is $1/n$. Thus

$$\varphi_1[v] = 1/n \sum_{k=0}^{n-1} \sqrt{k}$$

Now, for small n, this value may be computed directly. For larger values of n, it may be shown that this expression is approximately given by

$$\varphi_1[v] \approx [2(n-0.5)^{3/2} - 0.707] / 3n$$

For example, using $n = 50$ (so that there are 49 peasants), we see that $v(N) = 7$. Of these 7 units, the above formula gives us $\varphi_1[v] \approx 4.639$, and we see that the value gives the landlord well over half the total utility. The peasants would expect to get approximately $(7-4.639)/49$, or 0.0482 units each.

For very large values of n, the landlord should get about two thirds of the total utility, with the peasants sharing the remaining third, so that each peasant gets approximately $1/3\sqrt{n}$.

We may compare this with the "competitive market" solution of classical economics, according to which each peasant is paid an amount equal to the marginal productivity of one peasant, given that all peasants are in the coalition. This will be $\sqrt{(n-1)} - \sqrt{(n-2)}$, and for large n, this is approximately equal to $1/2\sqrt{n}$. Note that we obtain very different results according to the two schemes.

PROBLEMS ON THE SHAPLEY VALUE

1. Let the 4-person game v be given by

$$v(\{i\}) = 0 \text{ for all } i$$
$$v(\{1,2\}) = v(\{1,3\}) = 10$$
$$v(\{1,4\}) = v(\{2,3\}) = 50$$
$$v(\{2,4\}) = v(\{1,2,3\}) = 70$$
$$v(\{3,4\}) = 100$$
$$v(\{1,2,4\}) = 80$$
$$v(\{1,3,4\}) = v(\{2,3,4\}) = v(\{1,2,3,4\}) = 120.$$

Find its Shapley value.

2. For a 5-person game, define the vector

$$\mathbf{p} = (1, 3, 2, 6, 3)$$

and let v be given by

$$v(S) = \{\sum_{i \in S} p_i\}^2$$

for all $S \subset N$.

Find the Shapley value of this game, and show that it belongs to the core.

3. For each of the following 3-person games, assume $v(\{i\}) = 0$ for all i in n, and $v(N) = 100$. Find the Shapley value if:
   (a) $v(\{1,2\}) = v(\{1,3\}) = v(\{2,3\}) = 75$
   (b) $v(\{1,2\}) = v(\{1,3\}) = v(\{2,3\}) = 30$
   (c) $v(\{1,2\}) = 50$, $v(\{1,3\}) = 70$, $v(\{2,3\}) = 90$
   (d) $v(\{1,2\}) = 70$, $v(\{1,3\}) = 90$, $v(\{2,3\}) = 50$
   (e) $v(\{1,2\}) = 60$, $v(\{1,3\}) = 70$, $v(\{2,3\}) = 90$
   (f) $v(\{1,2\}) = 50$, $v(\{1,3\}) = 90$, $v(\{2,3\}) = 100$
   (g) $v(\{1,2\}) = 90$, $v(\{1,3\}) = 10$, $v(\{2,3\}) = 40$

4. VOTING STRUCTURES

We mentioned above that the formula (5.3.5) for the Shapley value is normally difficult to apply unless the game has some special structure. A type of game for which these calculations are considerably simplified is the simple game.

V.4.1. Definition. An n-person game, v, is said to be *simple* if, for every coalition S, $v(S)$ is either 0 or 1.

For a simple game, every coalition is either *winning* (which means, roughly, that it can obtain any imputation on which its members agree) or *losing* (which means that it can get its members nothing at all). If S is winning, then $v(S) = 1$, while if S is losing, then $v(S) = 0$.

For simple games, (6.13.5) is simplified by the observation that many of the terms

$$v(S\cup\{i\}) - v(S)$$

that appear in the sum will vanish, while the others are equal to 1. In fact, each of $v(S\cup\{i\})$ and $v(S)$ must be either 0 or 1. The difference will be 0 if S and $S\cup\{i\}$ are both winning coalitions, or both losing coalitions. It will be 1 if S loses but $S\cup\{i\}$ wins. (The remaining case, S winning but $S\cup\{i\}$ losing, is impossible by superadditivity.)

We find, then, that for simple games, the Shapley value can be given in the simplified form

5.4.1 $$\varphi_i[v] = \sum \gamma_{n,s}$$

where the summation is taken over all coalitions S such that S loses but $S\cup\{i\}$ wins.

An alternative method is to consider our probabilistic model once again. In an ordering of the players, we shall say that i is the pivot if the players preceding i in that ordering for a losing coalition S, but $S\cup\{i\}$ is a winning coalition. In other words, the *pivotal player* is the player that completes a winning coalition. It is not too difficult to see that, in a simple game, every ordering of the players has one and only on pivot. Then the power index φ_i is the probability that i will be pivotal, given that every ordering has probability $1/n!$.

Of great importance among simple games are *voting* games. In these, each player has a certain number of votes; a winning coalition is one with at least q votes, where q is often (but not always) equal to one more than half the total number of votes.

V.4.2. Example. Consider a corporation with four stockholders, holding, respectively, 1, 2, 3, and 4 shares of stock. Decisions can be made by any coalition having a simple majority of the total stock.

Since there are 10 shares of stock, 6 shares constitute a majority. We treat this as a voting game by assuming that the winning coalitions are those with 6 or more shares of stock. To compute the value, we consider all $4! = 24$ orderings of the 4 players:

```
123*4    213*4    312*4    412*3
124*3    214*3    314*2    413*2
132*4    231*4    321*4    42*13
134*2    234*1    324*1    42*31
142*3    241*3    34*12    43*12
143*2    24*31    34*12    43*12
```

In each ordering, we have designated the pivot by an asterisk. We see that, of the 24 orderings, player 1 is the pivot in 2 cases; players 2 and 3, in 6 cases each, and player 4, in 10 cases. Dividing by 24, we obtain the value vector

$$\varphi = (1/12, 3/12, 3/12, 5/12)$$

Note that the power indices of players 1 and 3 are slightly less than proportional to their shares of stock, while those of players 2 and 4 are slightly more than proportional. Note also that 2 and 3 have equal power, even though 3 has more stock than 2.

V.4.3. Example. The Security Council of the United Nations consists of 15 members, including 5 permanent members (the "Great Powers"). A motion is carried by vote of 9 members, except that opposition (veto) by any one of the permanent members defeats a motion.

Disregarding the possibility of abstention (used by the permanent members when they wish to show partial opposition to a motion), the winning coalitions in this game are all coalitions with at least nine members, including all of the permanent members. We shall compute the power index by considering one of the non-permanent members.

If i is a non-permanent member, the losing sets S such that $S \cup \{i\}$ is winning must consist of all five non-permanent members and exactly three non-permanent members. By (5.4.1), we must add the coefficients $\gamma_{n,s}$ for all these sets. For each of these, $s = 8$ and $n = 14$, so

$$\gamma_{n,s} = 8!5!/14! = 1/18018$$

We must multiply this by the number of such sets. The five permanent members can be chosen in only one way, but the three non-permanent members can be chosen from the remaining nine members in $_9C_3 = 84$ different ways. Thus

$$\varphi_i = 84/18018 = 0.00466$$

Thus, each non-permanent member has not quite ½ of 1 per cent of the total voting power. The ten non-permanent members together have 0.0466 of the voting power. The permanent members have the remaining 0.9534, or 0.1907 each. We see that the veto gives the permanent members almost complete control of the Security Council: each of them has more than 40 times as much power as each of the non-permanent members.

V.4.4. Example. A committee consists of four members, each with one vote. The chairman of the committee is given the power to break ties.

Considering all possible orders, we see that the chairman will be the pivot if he appears in second or third position. The probability of this is 2/4, or 1/2. Thus, letting player 1 be the chairman,

$$\varphi_1 = \tfrac{1}{2}$$

The remaining ½ of the power is divided equally among the three other members. Thus

$$\varphi_2 = \varphi_3 = \varphi_4 = 1/6$$

PROBLEMS ON VOTING STRUCTURES

1. A committee consists of a chairman with three votes, two senior members with two votes each, and five junior members with one vote each. Seven votes are needed to carry a motion. What is the distribution of voting power?

2. Suppose the committee of Example V.4.4 were changed so that the chairman had veto power. How would that change the distribution of power in the committee?

3. In a committee of five members, a simple majority of three is needed to carry a motion. Two members agree to join forces and vote together at all times. Does this strengthen or weaken them?

4. In a committee of five members, a unanimous vote is necessary to carry a motion. Two members agree to join forces and vote together at all times. Does this strengthen or weaken them?

5. Consider a three-person simple game, v, in which the one-person coalitions lose, but the two- and three-person coalitions win, i.e. $v(S) = 0$ if $|S| = 0$ or 1, but $v(S) = 1$ if $|S| = 2$ or 3. Show that the set **V** of three imputations

$$\mathbf{V} = \begin{Bmatrix} (½, ½, 0) \\ (½, 0, ½) \\ (0, ½, ½) \end{Bmatrix}$$

is *stable* in the following sense:
   (i) No imputation in this set dominates another (in the set).
   (ii) If **y** is any imputation not in **V**, then there is at least one **x** in the set **V** such that **x** dominates **y**. (See exercise 1 of Section 12, above, for domination.)
Note: sets having these two properties are generally known as stable set solutions or, alternatively, von Neumann-Morgenstern solutions.

6. In the game of Problem 5, above, suppose coalition {1,2} has formed. Prove that the imputation **x** = (½, ½, 0) is stable for this coalition in the following sense:
   (a) Players 1 and 2 together receive $v(\{1,2\})$
   (b) If either member of {1,2} wishes to obtain more, and defects to join player 3, then the remaining player can "Protect" his share of **x** while outbidding his former partner for player 3's services.

Show that **x** = (½, ½, 0) is the only imputation which is stable for {1,2} in this sense. Show, e.g. that the vector **y** = (0.49, 0.51, 0) is not stable because player 1 can form a coalition with 3 while paying 3 enough that 2 would be unable to obtain 3's services while paying 3 more, while protecting 2'2 share of 0.51.

(Payoff vectors which are stable in this sense are said to belong to the *bargaining set*.)

VI. DYNAMIC PROGRAMMING

1. THE PRINCIPLE OF MAXIMALITY

Suppose that a firm has a finite amount of resources, X, which is to be invested in n enterprises. It is assumed that these enterprises are all independent, so that the return from any one of them depends only on the resources invested in that one. If the return from an investment of x units in the ith enterprise is $g_i(x)$, the firm is faced with the problem of finding the function

6.1.1
$$F(x) = \max \sum_{i=1}^{n} g_i(x_i)$$

where the maximum is taken over all vectors $(x_1, x_2, \ldots, x_n)$ satisfying

6.1.2
$$\sum_{i=1}^{n} x_i = X$$

6.1.3
$$x_i \geq 0$$

We wish, then, to maximize the function

6.1.4
$$G(x_1, x_2, \ldots, x_n) = \sum_{i=1}^{n} g_i(x_i)$$

subject to the constraints (6.1.2) and (6.1.3). We do this by a technique which, while quite straightforward, yields very valuable (and unintuitive) results.

The *principle of maximality*, as it is called, is, on the face of it, nothing more than common sense. Yet even common sense often needs to be well formulated before it can be profitably used; the formulation of this principle must be recognized as an important step. It is generally attributed to R. Bellman, though a very similar technique was independently developed, more or less at the same time, by R. Isaacs.

To obtain a better understanding of the principle, we might think of our hypothetical firm's problem as a *sequential decision* problem. Let us assume that the firm must first decide how much of its resources to devote to the first enterprise. After making this decision, it must decide how much of the *remaining* resources to devote to the second enterprise,; then, how much of the remainder to devote to the third enterprise, and so on. This is probably easiest to visualize if we think of these decisions as being made in a different period of time. For instance, a firm might want to decide how to distribute its yearly advertising budget over the twelve months of the year.

We shall call any distribution of the resources over the n enterprises, a *policy*. It will be optimal if it maximizes the objective function (6.1.4) subject to (6.1.2) and (6.1.3). By a *sub-policy*, we shall mean a distribution of the resources left after k decisions, among the remaining $n-k$ enterprises. The *principle* of maximality can then be expressed in the intuitive form

6.1.5   *Every optimal policy contains only optimal sub-policies.*

In effect, (6.1.5) says that, whatever resources may have been spent on the first k enterprises, the remaining resources must be distributed optimally among the remaining $n-k$ enterprises. As such, its truth is obvious. Its importance, however, lies in the fact that the decision made at any one stage need consider only what can be done in subsequent stages with the amount of resources left; it need not be concerned with the allocation of resources already spent.

To see just what this means, let us return to the firm with a fixed yearly advertising budget, deciding how much to spend in each month of the year. When December comes, there will be no decision to make since all remaining monies are to be spent then: there is no question of saving resources for the future. In November the advertising manager must make a decision, but this is a relatively easy decision, as it is only a question of deciding how much to spend then and how much to save for December, and this process involves only maximizing a function of one variable. Then the October problem reduces to deciding how much to spend then and how much to solve for the combined November-December period. We find that the advertising director need make eleven easy (one-variable) decisions instead of one extremely complicated (11-variable) decision.

Let us give this a mathematical treatment. We shall use the notation $F_k(y)$ to denote the return that can be obtained by an optimal distribution of y units of the resource among the kth, $(k+1)$th, ..., and nth enterprises, i.e. among all enterprises from the kth one to the end.

Now the principle of optimality tells us that, whatever amount, x, may be invested in the kth enterprise, the remaining $y-x$ units must be invested optimally among the remaining $n-k$ enterprises. Thus,

6.1.6 $$F_k(y) = \max \{ g_k(x) + F_{k+1}(y-x)\}$$

where the maximum is taken over all x such that $0 \leq x \leq y$.

This will allow us to compute F_k if we know both the functions $g_k(x)$ and F_{k+1}. For the final decision, we have

6.1.7 $$F_n(y) = g_n(y)$$

The two equations, (6.1.6) and (6.1.7), represent the solution to our problem. In fact, (6.1.7) gives us the function F_n. This allows us (from (6.1.6)) to compute F_{n-1}, which in turn gives us F_{n-2}. Working backwards, this will give us all the functions F_k, all the way back to $k=1$. This tabulation should also include the optimal decision to be made at each stage for each level of resources then remaining.

There is a question yet as to the tabulation of these functions $F_k(y)$. In some cases, these will be elementary functions (powers, exponentials, logarithms, and such). Such cases are, unfortunately, rather rare: even if the g_k are elementary, it does not follow that the F_k will also be so. Rather, the F_k are generally made up of pieces of several elementary functions which can only be obtained in tabular form (see Figure VI.1.1.) Since an approximate solution is what is usually desired, this is not a great hardship; approximations can be made to as great a degree of accuracy as may be desired. In some cases the resources available may be made up of indivisible units; i.e. the x_k have to be integers. In this case, the dynamic programming technique is extremely powerful.

VI.1.1. Example. A company has 14 men whom it can assign to four projects. It estimates that the return on the ith project, given that x men are assigned to it, is the function $g_i(x)$ given in Table VI.1.1.

FIGURE VI.1.1 The functions $F_k(x)$ generally consist of pieces from several elementary curves.

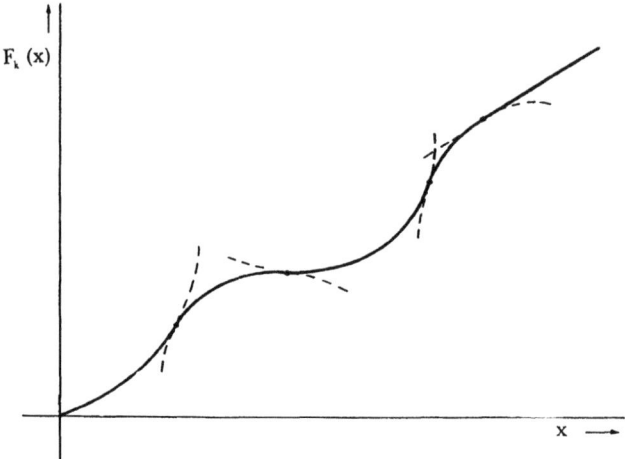

FIGURE VI.1.2 Returns on four jobs as a function of men assigned to each (Example VI.1.1).

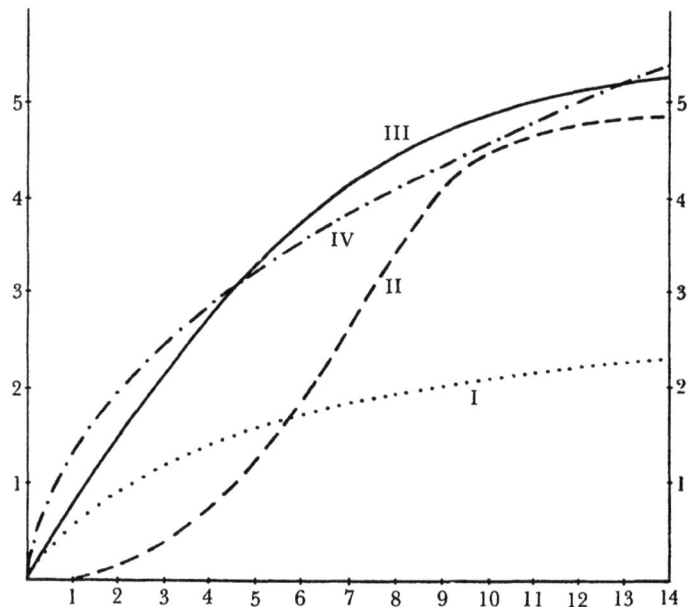

TABLE VI.1.1

x	i			
	1	2	3	4
0	0	0	0	0
1	0.53	0.04	0.75	1.35
2	0.88	0.18	1.50	1.98
3	1.15	0.40	2.15	2.48
4	1.37	0.75	2.75	2.88
5	1.58	1.25	3.30	3.22
6	1.75	1.90	3.75	3.56
7	1.86	2.64	4.15	3.87
8	1.95	3.43	4.48	4.13
9	2.04	4.12	4.73	4.37
10	2.13	4.50	4.90	4.62
11	2.20	4.70	5.05	4.85
12	2.25	4.82	5.15	5.05
13	2.29	4.89	5.22	5.25
14	2.33	4.90	5.26	5.45

In view of the nature of the problem, we can assume that only integer values are permissible for the variables. We must compute and tabulate the several functions F_k. We already know F_4: it is equal to g_4. To compute F_3, we have the definition

$$F_3(y) = \max_{0 \leq x \leq y} [g_3(x) + + F_4(y-x)]$$

We will simultaneously tabulate the function $x_3^*(y)$, defined by

$$F_3(y) = g_3(x^*) + F_4(y-x^*)$$

i.e., $x^*(y)$ is the number of men that should be assigned to the third enterprise, assuming that there are y men available for the third and fourth enterprises together.

To compute $F_3(14)$, say, we consider all the sums $g_3(x) + + F_4(14-x)$, for $0 \leq x \leq 14$. These are:

$$\begin{align}
g_3(0) + F_4(14) &= 0 + 5.45 = 5.45 \\
g_3(1) + F_4(13) &= 0.75 + 5.25 = 6.00 \\
g_3(2) + F_4(12) &= 1.50 + 5.05 = 6.55 \\
g_3(3) + F_4(11) &= 2.15 + 4.85 = 7.00 \\
g_3(4) + F_4(10) &= 2.75 + 4.62 = 7.37 \\
g_3(5) + F_4(9) &= 3.30 + 4.37 = 7.67 \\
g_3(6) + F_4(8) &= 3.75 + 4.13 = 7.88 \\
g_3(7) + F_4(7) &= 4.15 + 3.87 = 8.02 \\
g_3(8) + F_4(6) &= 4.48 + 3.56 = 8.04 \\
g_3(9) + F_4(5) &= 4.73 + 3.22 = 7.95 \\
g_3(10) + F_4(4) &= 4.90 + 2.88 = 7.78 \\
g_3(11) + F_4(3) &= 5.05 + 2.48 = 7.53 \\
g_3(12) + F_4(2) &= 5.15 + 1.98 = 7.13 \\
g_3(13) + F_4(1) &= 5.22 + 1.35 = 6.57 \\
g_3(14) + F_4(0) &= 5.26 + 0 = 5.26
\end{align}$$

The largest of these sums is 8.04, obtained with $x = 8$. We thus have

$$x_3^*(14) = 8 \qquad F_3(14) = 8.04$$

which means that, if there are 14 men available for projects 3 and 4 only, 8 should be assigned to project 3 and 6 to project 4, giving a total return of 8.04 units.

To determine $F_3(13)$, we consider all the sums $g_3(x) + F_4(13-x)$, for x from 0 to 13. These are, respectively, 5.25, 5.80, 6.35, 6.77, 7.12, 7.43, 7.62, 7.71, 7.70, 7.61, 7.38, 7.03, 6.30, 5.22. The largest of these is 7.71, obtained for $x = 8$. Therefore,

$$x_3^*(13) = 8 \qquad F_3(13) = 7.71$$

Continuing in this way, we complete the tabulation of $F_3(y)$, obtaining

TABLE VI.1.2

y	$x_3^*(y)$	$F_3(y)$
0	0	0
1	0	1.35
2	1	2.10
3	2	2.85
4	3	3.50
5	3	4.13
6	4	4.73
7	5	5.28
8	6	5.73
9	6	6.23
10	6	6.63
11	7	7.03
12	7	7.37
13	8	7.70
14	8	8.04

We are now in a position to compute $F_2(y)$, using the formula

$$F_2(y) = \max_{0 \le x \le y} [g_2(x) + + F_3(y-x)]$$

as well as $x_2^*(y)$, defined by

$$F_2(y) = g_2(x^*) + F_3(y-x^*).$$

Proceeding as before, we consider all the sums $g_2(x) + F_3(14-x)$, as x goes from 0 to 14. These are, respectively, 8.04, 7.74, 7.55, 7.43, 7.38, 7.48, 7.63, 7.92, 8.16, 8.25, 8.00, 7.55, 6.92, 6.24, 4.90. The largest of these is 8.25, obtained for $x = 9$. Therefore,

$$x_2^*(14) = 9 \qquad F_2(14) = 8.25$$

Continuing in this manner, we obtain x_2^* and F_2, given in

TABLE VI.1.3

y	$x_2^*(y)$	$F_2(y)$
0	0	0
1	0	1.35
2	0	2.10
3	0	2.85
4	0	3.50
5	0	4.13
6	0	4.73
7	0	5.28
8	0	5.73
9	0	6.23
10	0	6.63
11	0	7.03
12	0	7.37
13	0	7.70
14	9	8.25

This says that no men should be assigned to project 2 unless there are 14 men available for projects 2, 3, and 4 together. In this case, 9 men should be assigned to project 2.

We will now get the solution to the original problem. In fact, we are looking for $x_1^*(14)$ and $F_1(14)$. Again, we consider all the sums $g_1(x) + F_2(14-x)$, and find that these are 8.25, 8.23, 8.25, 8.18, 8.00, 7.81, 7.48, 7.14, 6.68, 6.17, 5.63, 5.05, 4.35, 3.63, and 2.33. Here is a tie for the maximum, with $x_1^*(14) = 0$ or $x_1^*(14) = 2$ both giving the maximum return of 8.25, and we see that there will be two optimal assignments of men. One is obtained by choosing $x_1 = 0$. There will then be 14 men left for projects 2, 3, and 4. We note that $x_2^*(14) = 9$, so we assign 9 men to project 2, leaving 5 for projects 3 and 4. Note next that $x_3^*(5) = 3$, and so we assign 3 men to project 3, leaving 2 men for project 4.

The other optimal assignment is obtained with $x_1^*(14) = 2$. If we assign 2 men to project 1, there are 12 left for projects 2, 3, and 4. Then we note that $x_2^*(12) = 0$ and $x_3^*(12) = 7$. This assigns no men to project 2, 7 to project 3, and 5 to project 4. It may be directly checked that these two assignments,

$$(0, 9, 3, 2)$$

and

$$(2, 0, 7, 5)$$

both give the total 8.25.

Example VI.1.1 shows the general procedure to be followed for dynamic programming problems. The computation gives us, not only the payoff functions $F_k(y)$, but also the assignment functions $x_k^*(y)$ which tell us the amount of resources to be assigned to the kth enterprise, assuming there are y units still available, i.e., $x_k^*(y)$ is defined by

6.1.8 $$F_k(y) = g_k(x_k^*) + F_{k+1}(y-x_k^*).$$

It is of interest to see how this technique compares with the crude evaluation of all possible assignments of the 14 men to four jobs. In the dynamic programming method, we notice that 15 additions were necessary to compute $F_3(14)$, 14 additions were necessary to compute $F_3(13)$, and so on. Thus we needed $1+2+\ldots+15 = 120$ additions to compute the entire function F_3. Another 120 were needed to compute the function F_2, and 15 were necessary to compute $F_1(14)$. In all, 255 additions were necessary. Against this, the total number of possible assignments is equal to the binomial coefficient $_{17}C_4$, or 680. A very real saving has been effected. More important, however, is the fact that, if a fifth project were available, the number of additions in the dynamic programming technique would increase only by another 120, to 375, while the total number of possible assignments would increase to $_{18}C_5$, or 3060. Clearly, the larger the problem, the greater the saving effected by dynamic programming will be. Although the dynamic program increases in an arithmetic progression, the number of possible assignments increases much more rapidly.

2. THE FIXED-CHARGE TRANSPORTATION PROBLEM

We consider here a problem similar to those treated in Chapter II, Section 11. Again, a company must deliver goods from its warehouses to its distributors. The difference lies in the fact that the cost function is not entirely linear. We shall assume that the cost of delivering x units from warehouse i to distributor j is $c_{ij}x$ (a linear cost), *plus* an additional amount k_{ij} which must be paid *if anything at all* is shipped from W_i to D_j. (For example, k_{ij} might be the cost of having a road repaired so that the company's trucks can use it.) This is the *fixed charge*: it is the same whether 1 unit or 1000 units are sent along this route.

Because the form of the cost function is not linear, we cannot solve this problem by linear programming techniques. However, if there are only *two* warehouses (or only two destinations), the dynamic programming technique can be used.

VI.2.1 Example. To see how this is done, let us assume that the availabilities at the two warehouses are, respectively, 9 and 13 units. We shall assume 6 destinations, with requirement vector

$$\mathbf{b} = (4, 3, 5, 4, 2, 4)$$

and cost matrices

$$C = \begin{bmatrix} 7 & 6 & 7 & 9 & 1 & 4 \\ 7 & 10 & 3 & 4 & 5 & 4 \end{bmatrix}$$

and

$$K = \begin{bmatrix} 7 & 7 & 1 & 0 & 4 & 9 \\ 3 & 1 & 6 & 6 & 0 & 4 \end{bmatrix}$$

We repeat that C is the matrix of linear costs, and K is that of fixed charges, i.e. the cost of shipping x units on the route (i, j) is $c_{ij}x + k_{ij}$ if x is positive, but 0 if $x = 0$.

Let us assume that x units of the good are shipped from the first warehouse to the jth destination. Since there is a total requirement of b_j units here, the remaining $b_j - x$ units must be sent from the second warehouse. The total cost of servicing this destination will be

6.2.1 $\quad g_j(x) = \begin{cases} c_{2j}b_j + k_{2j} & \text{if } x = 0 \\ c_{1j}b_j + k_{1j} & \text{if } x = b_j \\ c_{1j}x + c_{2j}(b_j - x) + k_{1j} + k_{2j} & \text{if } 0 < x < b_j \end{cases}$

The function $g_j(x)$ can be thought of as undefined (or very large) for $x < 0$ or $x > b_j$. The problem now becomes one of finding the minimum of

$$\sum_{j=1}^{n} g_j(x_j)$$

subject to the constraints that the x_j must be non-negative, and their sum equal to the total amount available at W_1. There are additional constraints, namely that the x_j cannot exceed the requirements b_j. We shall see, however, that these requirements present no problem at all: we simply ignore larger values of x_j in computing the objective functions F_k.

Since this is a minimization problem, we will have the equations

$$F_k(y) = \min \{g_k(x) + F_{k+1}(y-x)\}$$

Where the minimum is taken over all non-negative x such that $x \leq y$ and $x \leq b_k$.

The first problem is to tabulate all the functions $g_k(x)$. This is not, strictly speaking, necessary, but it helps to see things at a glance. It can be proved (though we omit the proof here) that the minimum for a problem such as this will always be obtained at an extreme point of the constraint set. We saw in Chapter III that, for problems such as this, the extreme points have only integer values for the variables (assuming, of course, that the requirements and availabilities are themselves integers). Thus we need only tabulate the functions, given by (6.2.1), for integer values of x. These are given in

TABLE VI.2.1

x	\multicolumn{6}{c}{j}					
	1	2	3	4	5	6
0	31	31	21	22	10	20
1	38	34	26	27	10	29
2	38	30	30	32	6	29
3	38	25	34	37	--	29
4	35	--	38	36	--	25
5	--	--	36	--	--	--

Note that the functions are simply not defined for $x > b_j$.

As before, we have $F_6 = g_6$. This allows us to compute $F_5(y)$, as well as $x_5^*(y)$, for y between 0 and 6 (Table VI.2.2).

TABLE VI.2.2

y	$x_5^*(y)$	$F_5(y)$
0	0	30
1	1	30
2	2	26
3	2	35
4	2	35
5	2	35
6	2	31

We can, now, compute $F_4(y)$ and $x_4^*(y)$ for y between 0 and 9. Note that 9 is the total availability at W_1; hence we need not consider larger values of y.

TABLE VI.2.3

y	$x_4^*(y)$	$F_4(y)$
0	0	52
1	0	52
2	0	48
3	1	53
4	0	57
5	0	57
6	0	53
7	1	58
8	2	63
9	3	67

Continuing, we now compute $F_3(y)$ and $x_3^*(y)$

TABLE VI.2.4

y	$x_3^*(y)$	$F_3(y)$
0	0	73
1	0	73
2	0	69
3	1	74
4	2	78
5	0	78
6	0	74
7	0	79
8	2	83
9	3	87

Next, we get $F_2(y)$ and $x_2^*(y)$

TABLE VI.2.4

y	$x_2^*(y)$	$F_2(y)$
0	0	104
1	0	104
2	0	100
3	3	98
4	3	98
5	3	94
6	3	99
7	3	103
8	3	103
9	3	99

Finally, using this, we find

$$x_1^*(9) = 4; \quad F_2(9) = 129$$

This gives us the solution: we must let $x_1 = 4$. This leaves 5 units at W_1; we see next that $x_2^*(5) = 3$. Then $x_3^*(2) = 0$, $x_4^*(2) = 0$, and $x_5^*(2) = 2$. Thus the amounts shipped from W_1 are given by the vector

$$(4, 3, 0, 0, 2, 0)$$

The total shipping schedule is obtained by including also the amounts shipped from W_2. This gives us the matrix

$$X = \begin{bmatrix} 4 & 3 & 0 & 0 & 2 & 0 \\ 0 & 0 & 5 & 4 & 0 & 4 \end{bmatrix}$$

It may be checked that this will indeed give a total cost of 129 units. Note also that the program is degenerate: this degeneracy, which causes so many problems for linear programs, has no effect all in the dynamic programming computations.

PROBLEMS ON DYNAMIC PROGRAMMING

1. A company has 10 men which it must assign to four different projects. The revenue expected from assigning j men to the ith project is given in the following table

Men Assigned	Job			
	1	2	3	4
0	0	0	0	0
1	13	8	15	10
2	23	17	22	20
3	29	29	31	30
4	37	31	39	35
5	42	40	44	40
6	51	47	51	50
7	64	54	62	65
8	80	67	70	78
9	88	80	80	88
10	95	87	91	99

How many men should be assigned to each job, so as to maximize expected revenue?

2. A company wishes to send a cargo of gadgets, widgets, and beepers to a distributor; it has a truck with a capacity of 25,000 lb. Each gadget weights 3000 lb., each widget 5000 lb., and each beeper 4000 lb. The demand for these items is a random variable; the probability that there be a demand for k items of a given type is given in the following table:

	Gadgets	Widgets	Beepers
0	0.05	0.05	0
1	0.15	0.10	0.05
2	0.15	0.15	0.08
3	0.15	0.20	0.11
4	0.15	0.15	0.20
5	0.10	0.15	0.15
6	0.10	0.10	0.13
7	0.10	0.05	0.10
8 or more	0.05	0	0.18

How many items of each type should be sent so as to maximize expected profits? It may be assumed that leftover items can be kept at very little cost, so that they represent no loss.

3. The following are fixed-cost transportation problems. In each case, **a** is the vector of availabilities at the two sources, and **b** is the vector of requirements at the n destinations. The matrices C and K represent, respectively, the linear costs per unit between the sources and destinations, and the fixed costs of keeping the corresponding routes open.

(a)   **a** = (8, 13)   **b** = (3, 3, 5, 4, 6)

$$C = \begin{bmatrix} 3 & 5 & 2 & 6 & 1 \\ 1 & 3 & 1 & 8 & 2 \end{bmatrix} \qquad K = \begin{bmatrix} 1 & 4 & 8 & 3 & 4 \\ 4 & 5 & 6 & 2 & 3 \end{bmatrix}$$

(b)   **a** = (13, 17)   **b** = (5, 7, 8, 3, 2, 5)

$$C = \begin{bmatrix} 6 & 1 & 3 & 5 & 8 & 2 \\ 2 & 2 & 4 & 6 & 3 & 1 \end{bmatrix} \qquad K = \begin{bmatrix} 1 & 4 & 2 & 3 & 0 & 5 \\ 5 & 3 & 2 & 8 & 3 & 1 \end{bmatrix}$$

(c)   **a** = (9, 6)   **b** = (3, 4, 5, 3)

$$C = \begin{bmatrix} 1 & 2 & 6 & 4 \\ 2 & 0 & 4 & 4 \end{bmatrix} \qquad K = \begin{bmatrix} 2 & 3 & 1 & 6 \\ 0 & 2 & 5 & 3 \end{bmatrix}$$

3. INVENTORIES

An important application of mathematical analysis is to inventory control. Generally speaking, an inventory is a supply of goods which is kept in storage to service a future demand. The decision-maker will, in normal situations, wish to maintain a certain amount of inventory, in part because of uncertainties in demand, and in part because the "setting-up" costs connected with keeping supply exactly equal to demand at all costs can become prohibitive. On the other hand, too large an inventory ties up working capital and must, moreover, be stored (for a price) in some warehouse. An optimal policy must be sought to compromise between these two extremes.

Let us see how this is done. We assume that the demands for each month, over a period extending n months into the future, are known. (To simplify the problem, assume that the demand is effective at the end of each month.) Let d_j be the amount needed at the end of the jth month, x_j the amount produced during the month, and y_j the number of units in inventory at the *beginning* of the month. Then the x_j, y_j and d_j are connected by the equation

6.3.1 $$y_{j+1} = y_j + x_j - d_j$$

and we have, moreover, the constraints

6.3.2 $$x_j, y_j \geq 0.$$

The total costs connected with the inventory system include setting-up costs, production costs, and holding (storage) costs. We shall simplify the model by assuming that production runs can only be made once each month (though there is no need to make a run every month). The cost of producing x units in a month will be a function, $c_j(x)$, while the cost of keeping y units in inventory will be another function, $f_j(y)$. The total costs connected with the jth month will be

6.3.3 $$g_j(x_j, y_j) = K\delta_j + c_j(x_j) + f_j(y_j)$$

where K is the setting-up cost, and $\delta_j = 0$ if $x_j = 0$, but is equal to 1 if $x_j > 0$. Then the problem is to minimize

$$\sum_{j=1}^{n} g_j(x_j, y_j)$$

where g_j is as in (6.3.3), subject to the constraints (6.3.1)-(6.3.2). In form, this is slightly different from the problems studied in the first two sections of this chapter. Essentially, however, the problem is the same and can be solved by the method of dynamic programming.

Let us, as before, write

6.3.4
$$F_k(\xi) = \min \sum_{j=k}^{n} g_j(x_j, y_j)$$

where ξ represents the inventory on hand at the beginning of the kth period. The minimum is taken over all (x_j, y_j) satisfying (6.3.1) and (6.3.2), and such, moreover, that $y_k = \xi$.

If an amount x_k is produced during the kth period, then the inventory on hand at the end of this period is $\xi + x_k - d_k$. Applying the recurrence relation of the principle of maximality, we have

6.3.5
$$F_k(\xi) = \min \{ g_k(x, \xi) + F_{k+1}(\xi + x - d_k) \}$$

where the minimum is taken over all $x \geq 0$.

We shall use (6.3.5) to compute the functions F_k. It very often happens that the functions c_j and f_j are equal (i.e. do not depend on j). This will simplify the calculations considerably, and especially so if these functions are linear.

VI.3.1. Example. A tool company has orders for wrenches to be delivered over the course of a year. The total number to be delivered each month is shown in the table.

TABLE VI.3.1

Month	Requirement
January	300
February	100
March	200
April	400
May	300
June	300
July	500
August	700
September	400
October	700
November	200
December	600

We shall assume that each wrench costs $3.00 to produce, plus a setting-up cost of $100 per production rum. Holding costs are 10¢ per wrench per month. (This includes both the cost of storage and the interest costs of the money invested.)

The fact that the variable costs of production (as opposed to the setting-up costs) are linear means, more or less, that they can be disregarded. There are 4600 wrenches to be produced; these will cost $13,800 regardless of when they are produced. (This is a considerable simplification: there is no reason why costs should be linear, and moreover it may be that costs may be higher in some months that in other.) For this problem, therefore, we shall consider only the setting-up and holding costs.

We proceed, therefore, to calculate the functions g_j and F_k. Disregarding production costs, we have

$$g_j(x, y) = \begin{array}{ll} 0.1y & \text{if } x = 0 \\ 0.1y + 100 & \text{if } x > 0 \end{array}$$

for each j. Suppose, now, that we wish to end the year with no inventory on hand, i.e., $y_{13} = 0$. The month of December can be started with any number of wrenches up to 600 on hand. Of necessity, we will have

$$x_{12} = 600 - y_{12}$$

and so we can form a table (VI.3.2) of the functions $F_{12}(y)$, $x_{12}*(y)$.

TABLE VI.3.2

y	$x_{12}^*(y)$	$F_{12}(y)$
0	600	100
100	500	110
200	400	120
200	300	130
400	200	140
500	100	150
600	0	60

where, as before, $x_{12}^*(y)$ is the amount to be produced in December given an inventory level y at the beginning of the month.

We proceed, next, to calculate F_{11}. For example, in case $y_{11} = 0$, we set

$$F_{11}(0) = \min \{ g_{11}(x, 0) + F_{12}(x - 200) \}$$

since the requirement for November is 200 wrenches. It is easy to see that x must be at least 200, and that the desired value $F_{11}(0)$ is then the smallest of the numbers 200, 210, 220, 230, 240, 250, 160. The minimum, 160, is obtained by letting $x = 800$. Thus

$$x_{11}^*(0) = 800, \quad F_{11}(0) = 160.$$

We proceed, similarly, to compute the functions $x_k^*(y)$, $F_k(y)$, shown in the Table VI.3.3.

y	F_2	F_3	F_4	F_5	F_6	F_7	F_8	F_9	F_{10}	F_{11}	F_{12}
0	750	710	650	590	510	460	360	310	220	160	100
100	720	720	660	600	520	470	370	320	230	170	110
200	740	670	610	610	530	480	380	330	240	120	120
300	700	690	630	620	490	490	390	340	250	140	130
400	730	650	650	550	510	500	400	260	260	160	140
500	700	680	670	570	530	410	410	280	270	180	150
600	740	710	610	590	550	430	420	300	280	200	60
700	780	740	640	560	570	450	380	320	230	220	--
800	820	690	670	590	490	470	400	340	250	140	--
900	770	730	650	620	520	490	420	360	210	--	--
1000	830	770	690	650	550	510	440	380	240	--	--

y	x_2^*	x_3^*	x_4^*	x_5^*	x_6^*	x_7^*	x_8^*	x_9^*	x_{10}^*	x_{11}^*	x_{12}^*
0	500	400	600	700	800	500	1100	1300	900	800	600
100	0	300	500	600	700	400	1000	1200	800	700	500
200	0	0	0	500	600	300	900	1100	700	0	400
300	0	0	0	400	0	200	800	1000	600	0	300
400	0	0	0	0	0	100	700	0	500	0	200
500	0	0	0	0	0	0	600	0	400	0	100
600	0	0	0	0	0	0	500	0	300	0	0
700	0	0	0	0	0	0	0	0	0	0	--
800	0	0	0	0	0	0	0	0	0	0	--
900	0	0	0	0	0	0	0	0	0	--	--
1000	0	0	0	0	0	0	0	0	0	--	--

From this, we now calculate the optimal policy. In fact, the table allows us to compute that

$$x_1^*(0) = 600 \qquad F_1(0) = 800$$

The optimal policy is then given by the recursive definition

$$y_1(0) = 0$$
$$x_k = x_k^*(y_k)$$
$$y_{k+1} = y_k + x_k - d_k$$

and Table VI.3.4.

TABLE VI.3.4

k	1	2	3	4	5	6	7	8	9	10	11	12
y_k	0	300	200	0	400	0	500	0	400	0	200	0
x_k	600	0	0	600	0	800	0	1100	0	900	0	600

We see that six production runs are scheduled. Most of these are for two months' requirements, but the first is for three months, and the last is for one month. Note that no production is ever scheduled until inventory has vanished: this is a common property of systems such as this (though it might not be if the cost functions are more complicated). The reason is that it is clearly better to plan production so as to avoid any leftovers, which contribute nothing but expenses.

It can be checked directly that this policy gives a total of $200 for holding costs, plus $600 for setting-up costs. Added to the $13,800 already discussed, the total cost is $14,600.

VI.3.2. Example. A vulcanizing plant must produce 30 tires per month, during five months, to meet a contract. The costs of production and holding costs are independent of time, and given in Table VI.3.5.

TABLE VI.3.5

x (Tires)	Production cost	Holding cost (per month)
10	170	15
20	300	30
30	430	45
40	550	75
50	670	105
60	790	150
70	900	
80	1000	
90	1090	

We shall assume that there are no tires on hand at the beginning of the first month, and none on hand at the end of the last month. Proceeding as before, we obtain the functions $x_5^*(y)$ and $F_5(y)$, shown in

TABLE VI.3.6

y	$x_5^*(y)$	$F_5(y)$
0	30	430
10	20	315
20	10	200
30	0	45

And so on. We finally obtain VI.3.7 for x_k^* and F_k:

TABLE VI.3.7

y	x_1^*	x_2^*	x_3^*	x_4^*	x_5^*	F_1	F_2	F_3	F_4	F_5
0	30	60	30	60	30	2100	1670	1265	835	430
10	--	50	20	50	20	--	1565	1150	730	315
20	--	40	10	40	10	--	1460	1005	625	200
30	--	0	0	0	0	--	1310	880	475	45
40	--	0	0	0	0	--	1225	805	390	--
50	--	0	0	0	0	--	1110	730	420	--
60	--	0	0	0	0	--	1030	625	195	--

We conclude from this that the optimal policy is to produce 30 tires in the first month, and 60 each in the second and fourth months. The total cost is $2100.

PROBLEMS ON INVENTORIES

1. A company produces bolts; there is a steady demand for 160 T. of bolts per month. Production costs are $1250 per ton, plus $3000 in setting-up costs for each production run. Inventory holding costs are $200 per ton, per year. How frequently should production runs be scheduled to minimize total costs over the next eight months?

2. A canned food company has a steady demand for 5000 cans of beans per month. Production costs are 32¢ per can, plus $30 setting-up costs for each run. Inventory holding costs are 10¢ per can per year. How many cans should be made at each production run to minimize costs over the next six months?

3. A miller in a resort town has a contract to deliver 400 lb. of flour during the five-week season. Production and holding costs (per week) for the flour are given in the following table:

Pounds	Production	Storage
100	13	2
200	17	4
300	21	6
400	25	8
500	28.5	9.5
600	32	11
700	35.5	12
800	39	13
900	42	14
1000	45	14.5
1100	47.5	15
1200	50	15.5

How should the miller schedule his production?

4. A tool company has a contract to deliver 300 screwdrivers in August, 500 in September, 200 in October, 400 in November, and 600 in December. Production costs are 75¢ per screwdriver, plus $25 setting-up costs for each run. Holding costs are 3¢ per screwdriver per month. How should production be scheduled?

4. STOCHASTIC INVENTORY SYSTEMS

In the preceding section of this chapter, we assumed that the demands of the system were deterministic, i.e., not subject to random fluctuations. Planning can be done well in advance, and there is never any danger that the system will be unable to meet demands. In practice, demands are often known only probabilistically. Because of this, there is a very real danger that the stock will run out. This may cause a penalty cost. As a result, we shall see, prudence will almost always demand that such systems be maintained at a "safe" level: more stock is kept than one expects to use.

We shall, in what follows, make the simplifying assumption that a production run can be ordered at the beginning of any time period but at no other time. The amount produced in this period is immediately available; i.e. it can be used to meet demands received during this period. (This does not mean, of course, that production is instantaneous, but rather that there are substantially equal lags in the production and retail processes, i.e., customers are usually willing to wait a while before receiving their orders.) We assume also that the demand during the time period is a random variable with a known distribution.

In general, suppose there is an inventory y_j on hand at the beginning of the jth period. If we then order x_j units, we will have $x_j + y_j$ units available for this period. If the demand is Z_j units, we will be able to deliver either Z_j or $x_j + y_j$ units, whichever is smaller. Thus we will have

6.4.1 $\qquad y_{j+1} = x_j + y_j - Z_j \quad \text{if} \quad Z_j \leq x_j + y_j$
$\qquad\qquad\qquad\;\; 0 \qquad\qquad \text{if} \quad Z_j \geq x_j + y_j$

Consider next the costs incurred in the jth period. There is, first of all, the cost $f_j(y_j)$ of holding y_j units in inventory. Next, there are production costs $c_j(x_j)$, plus a setting-up cost $A\delta_j$, where $\delta_j = 0$ if $x_j = 0$, but $\delta_j = 1$ if $x_j > 0$.

Finally, a penalty cost p (per unit) is incurred whenever $Z_j > x_j + y_j$. For simplification, we shall assume that the charges are linear functions of their variables.

6.4.2 $\qquad g_j(x_j, y_j, Z_j) = cx_j + hy_j + A\delta_j \qquad\qquad\qquad\quad \text{if} \quad Z_j \leq x_j + y_j.$
$\qquad\qquad\qquad\qquad\quad\; cx_j + hy_j + A\delta_j + p(Z_j - x_j - y_j) \quad \text{if} \quad Z_j > x_j + y_j$

The problem, now, is to choose the variables, x_j, in such a way as to minimize the expected sum of the costs g_j over the n periods. We will have

$$F_k(\xi) = \min \mathrm{E}\left[\sum_{j=k}^{n} g_j(x_j, y_j, Z_j)\right]$$

where the minimum is taken over all non-negative $x_k, \ldots, x_n$, with y_j and g_j given by (6.4.1) and (6.4.2), and $y_k = \xi$. We will then have

6.4.3 $\qquad F_k(y_k) = \min_x \mathrm{E}\left[g_k(x, y_k, Z_k) + F_{k+1}(y_{k+1})\right]$

The usual dynamic programming technique will be used, but with expected values rather than deterministic costs.

VI.4.1. Example. A butcher must decide how many turkeys to keep in stock during a three-day period. He considers that he might sell as many as four turkeys each day, with the probability distributions (for each day) given in Table VII.4.1.

TABLE VI.4.1

Number of turkeys	Day		
	Thursday	Friday	Saturday
0	0.3	0.2	0.1
1	0.4	0.3	0.1
2	0.2	0.3	0.3
3	0.1	0.1	0.4
4	0	0.1	0.1

Each turkey costs $3, but there is a charge of $2 for each delivery from the farm (regardless of the number of turkeys in the delivery). The storage charge for leftover turkeys is 50¢ per turkey per night. Additionally, there is a penalty of $7 (in lost business and good will) for each customer who must be sent away for lack of a turkey.

Once again, the problem is to be solved by the "backward" technique. Suppose that the stock level on Saturday (including any deliveries made on Saturday morning) is three turkeys. In that case, there is a 0.1 probability of three leftover turkeys, causing costs of $1.50; a 0.1 probability of two leftover turkeys, causing costs of $1; a 0.3 probability of one leftover turkey, with a cost of 50¢; and a 0.1 probability that there will be a shortage of one turkey, causing a penalty of $7. The expected costs for the day, *exclusive of order costs*, amount to

$$(0.1)(1.5) + (0.1)(1) + (0.3)(0.5) + (0.1)(7) = 1.10$$

or $1.10. For other possible levels of supply $x_3 + y_3$, the expected inventory and penalty charges are computed and given in Table VI.4.2.

TABLE VI.4.2

$x_3 + y_3$	Costs
0	16.10
1	9.85
2	4.35
3	1.10
4	0.85
5	1.35
6	1.85
7	2.35
8	2.85
9	3.35
10	3.85

It is clear that costs on Saturday will be lowest if we have four turkeys available on that day. Suppose, however, that the stock on Friday night (at closing time, i.e., y_3) is less than four turkeys. Then the butcher must decide whether to order any for Saturday, remembering that order costs of $3 per bird, plus a $2 delivery charge, must then be paid. It may be seen, for example, that if $y_3 = 1$, then it is best to order 2 turkeys, incurring order charges of $8, but lowering the total expected costs from $9.85 to $9.10. If, on the other hand, $y_3 = 2$, then no order should be made since the costs of ordering would exceed any savings in future costs. More generally, it turns out that if $y_3 = 0$ or 1, an order should be made to increase the supply to three. If $y_3 \geq 2$, however, no order should be made. We obtain in this way the functions $x_3^*(y)$ and $F_3(y)$, as shown in Table VI.4.3.

TABLE VI.4.3

y	$x_3^*(y)$	$F_3(y)$
0	3	12.10
1	2	9.10
2	0	4.35
3	0	1.10
4	0	0.85
5	0	1.35
6	0	1.85
7	0	2.35
8	0	2.85
9	0	3.35
10	0	3.85

We repeat the procedure. Suppose we have no turkeys available on Friday morning. Then there is an expected shortage of 1.6 turkeys, with a corresponding expected penalty of $11.20, plus the prospect of having no birds on Friday night, which as we have just calculated, represents expected costs of $12.10. The total expected costs are then $23.30. In general, for various levels of supply $x_2 + y_2$, the expected future costs are given by

TABLE VI.4.4

$x_2 + y_2$	Costs
0	23.30
1	17.20
2	12.10
3	8.12
4	5.12
5	3.90
6	3.77
7	4.32
8	5.25
9	6.25
10	7.25

As before, we can compute the optimal policy for each level of supply on Thursday night (y_2): it will be seen that if $y_2 = 0$ or 1, then an order should be made, raising the level to either 3 or 4 (there is a tie here). Otherwise no order should be made. We have

TABLE VI.4.5

y	$x_2^*(y)$	$F_2(y)$
0	3 or 4	19.12
1	2 or 3	16.12
2	0	12.20
3	0	8.12
4	0	5.12
5	0	3.90
6	0	3.77
7	0	4.32
8	0	5.25
9	0	6.25
10	0	7.25

Finally, the expected costs, for each available Thursday morning supply $x_1 + y_1$, are given in Table VI.4.6.

TABLE VI.4.6

$x_1 + y_1$	Costs
0	27.82
1	21.17
2	17.04
3	13.40
4	10.29
5	7.01
6	6.97
7	7.05
8	9.00
9	10.20
10	10.68

From this, we can now find the optimal policy for a given y_1 (Wednesday night supply): if $y_1 = 0$ or 1, we place an order to raise the level to five birds; if $y_1 \geq 2$, no order is placed. The functions $x_1^*(y)$ and $F_1(y)$ are given in Table VII.4.7.

TABLE VI.4.7

y	$x_1^*(y)$	$F_1(y)$
0	5	24.01
1	5	21.01
2	0	17.04
3	0	13.40
4	0	10.29
5	0	7.01
6	0	6.97
7	0	7.05
8	0	9.00
9	0	10.20
10	0	10.68

We have thus determined the optimal policy for this butcher. This policy can best be characterized by pointing out that for each day k, there are two numbers, s_k and S_k. These numbers are (1,5), (1,4) and (1,3), respectively, for $k = 1, 2,$ and 3. If, for period k, the supply $y_k \leq s_k$, then an order should be made to raise the supply to S_k. If, on the other hand, $y_k > s_k$, no order should be made. Many inventory problems exhibit this type of solution. A sufficient condition for this type of behavior is that the costs (production costs, holding costs, and shortage penalties) all be linear functions of the relevant variables.

PROBLEMS ON STOCHASTIC INVENTORY SYSTEMS

1. An automobile dealer makes a profit of $800 on each car that he sells; his holding cost is $85 per car per month. The demand for this car is a random variable with distribution given by the following table.

Demand	Probability
0	0.05
1	0.15
2	0.23
3	0.23
4	0.17
5	0.11
6	0.04
7 or more	0.02

Assuming that his shortage penalty is simply the $800 (lost profit) and that he can order cars only at the beginning of each month, how many automobiles should he keep in stock?

2. A television dealer sells TV sets for $400. He pays storage costs of $10 per set per month; his supply costs are $300 per set plus $50 delivery charge for any shipment of whatever size. For the last three months of the model year, the demand is expected to have the distribution shown in the table.

	July	August	September
0	0.05	0.05	0.10
1	0.15	0.20	0.30
2	0.20	0.30	0.25
3	0.30	0.20	0.20
4	0.20	0.15	0.10
5	0.10	0.10	0.05

Sets can be ordered at the beginning of each month; if demand is higher than supply, the excess demand is simply lost. Any sets left over at the end of September must be remaindered for $100 each. What is the best ordering policy?

3. An airplane company has noticed that, in any month, the demand for its "family jet" model is a random variable with probability distribution $P(0) = 0.1$, $P(1) = 0.6$, $P(2) = 0.3$. Production costs are $50,000 per plane, plus a $10,000 setting-up cost for each production run. Holding costs are $4000 per plane per month. Each plane can be sold for $70,000, but any left over at the end of the model year must be sold below cost, at $40,000. Assuming that at most one production run can be scheduled in any month, how should the company schedule production for the 12-month model year?

VII. GRAPHS AND NETWORKS

1. INTRODUCTION

In this chapter we study mathematical systems known as *networks*. One special problem of this type, the *transportation problem*, was studied in some detail in Chapter II. Here we shall be interested in more general networks.

The theory of networks is an old and venerable branch of applied mathematics, dating back to the physicists James Clark Maxwell (1831-1879) and Gustav Kirchhoff (1824-1887) who were, however, interested in the study of electrical networks. The economically oriented theory of networks dates to the 1940's and 50's. Because many network problems can be treated by linear and dynamic programming, the authors mentioned in those chapters have clearly contributed to this subject.

In the context of this chapter (as in chapter III, Section 12), a graph is a collection of points, called *nodes*, together with lines, called *arcs*, that join pairs of these nodes. We assume that there is at most one arc joining any pair of nodes. The graph is *connected* if, given any two nodes, A and B, it is possible to go from A to B along a path made from the arcs of the graph. The graph in Figure VII.1.1 is connected, whereas that in Figure VII.1.2 is *disconnected*: it is not connected because it is impossible to go from A to G along the arcs of the graph.

FIGURE VII.1.1. A connected graph.

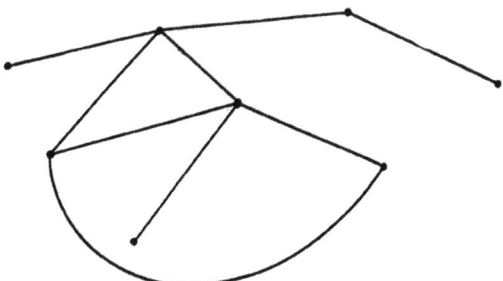

A network, as we shall use the word here, is a connected graph. Normally, each of the arcs in a network has some number attached to it, corresponding perhaps to the capacity of the arc, or to the distance between the two nodes that it joins. The network may be oriented, which means that each arc has a definite direction (say from A to B but not from B to A) or that the capacities or distances along the arc are different in thee two directions, or it may be *unoriented*. We shall find that it does not matter, for many applications, whether the graph is oriented or not. Where it does matter, we shall definitely say so.

FIGURE VII.1.2. A disconnected graph.

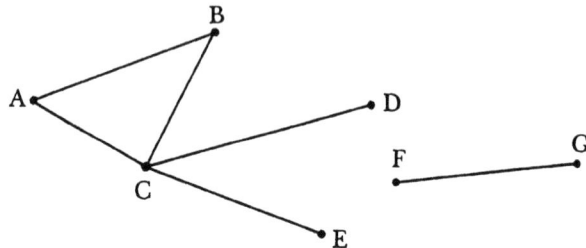

2. CRITICAL PATH ANALYSIS

The method of *critical path analysis* is another example of the ways in which "plain common sense" can be used for rather complex problems, if only it is correctly formulated. Using no advanced mathematics (though its problems could be reformulated as linear programs), critical path analysis is for small problems an obvious way to do things; many people have used it for years without thinking about it. For larger problems, it is not so common, perhaps because people do not take the time to formulate problems in these terms. The strict mathematical formulation of the method is principally the work of J. E. Kelley, Jr. and M. R. Walker, and dates back to the 1950's.

Critical path analysis is used to find the minimum time necessary to complete a project. It also shows which of the several parts of the project actually affect the overall completion time.

Generally speaking, we define a *project* as a set of *activities*, together with an *order relation* among these activities; we say that the ith activity precedes the jth activity if the former must be completed before the latter can begin. There is, furthermore, a nonnegative number, t, assigned to each activity, representing the time it takes to carry out this activity, the *time of performance* of the activity.

A project can be represented by a directed graph, whose nodes represent the several activities. There will be an arc from node i to node j if the ith activity precedes the jth activity. (Actually, it is not necessary to draw an arc from i to j unless j is an immediate successor of i.

VII.2.1. Example. Let us consider the project of repairing a broken door for an automobile. The door must be removed, hammered back into shape, and welded. Some parts must also be ordered. We have a table:

TABLE VII.2.1

Activity	Time (days)	Immediate predecessors
a. start	0	None
b. remove door	1	a
c. order parts	3	a
d. weld and paint door	2	b,c
e. reinstall window	1	b,c
f. replace door	1	d,e
g. finish	0	f

The project can be represented by the network shown in Figure VII.2.1. As may be seem, we assign to each arc of the network a number corresponding to the time of performance of the activity that is at the *beginning* of the arc. We call this the length of the arc. The network diagram is especially useful in displaying the order of work: one does not need to know much about automotive repair in order to understand it.

FIGURE VII.2.1. The graph for Example VII.2.1.

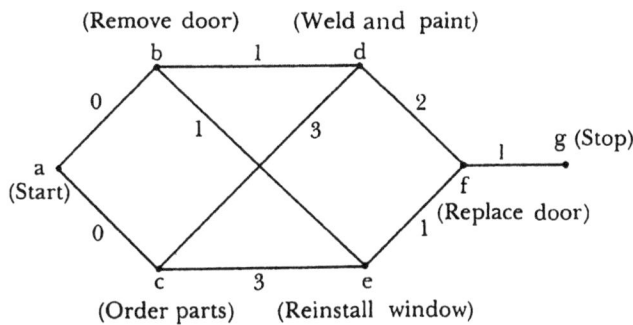

Suppose we wish to know the earliest possible finishing time for the project. This can be done in either of two ways. One is to treat the problem as a linear program and minimize the finishing time, subject to the obvious constraints (e.g., that activity d cannot begin until after activity b is finished). The other is to think of this as a dynamic program and look for the longest path through the network, from start to finish. We shall, in essence, do this.

Let us (fittingly) start at node *a*, the "start" of the network. Starting here, we assign times to each arc and node, as follows:
1. Node *a* is assigned the time 0.
2. Each arc is assigned a time equal to the sum of its length, and the time assigned to the node at which the arc begins.
3. Each node is assigned a time equal to the maximum of the times assigned to the arcs that begin at the node.

Let us see exactly what this does. The time assigned to each node is the earliest possible time at which the corresponding activity can be started, and the number assigned to an arc is the earliest possible time at which such an activity can be finished. Rule 1 tells us that we start at time 0; rule 2 tells us that the finish time for an activity is equal to its starting time plus the time it takes (the length of the arc); rule 3 tells us that an activity can be started as soon as its predecessors have been completed, but no sooner. The times obtained in this manner are known as the *early start time* (e.s.t.) and the *early finish time* (e.f.t.) of the various activities. The e.s.t. for the "finish" node is the earliest time at which the project can be finished and is known as the *target time* of the project. We shall use τ to denote this.

Figure VII.2.2 shows our network with e.s.t. and e.f.t. assigned to each of the activities. As may be seen, the target time is 6 (days).

FIGURE VII.2.2. Early start times (at nodes) and early finish times (on arcs) for Example VII.2.1.

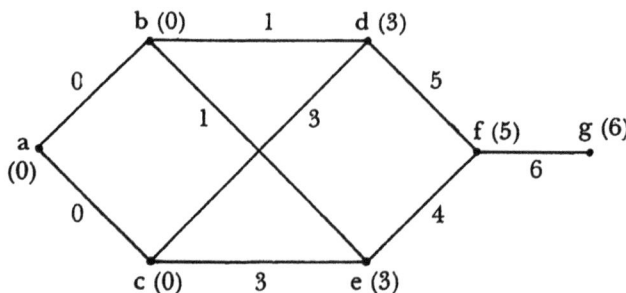

Suppose, next, that for some reason one of the activities is delayed. We would like to know whether this will affect the target time of the project.

What we really which to know is the latest time at which a given activity can be started without affecting the target time. We shall call this the late start time (l.s.t.) of the activity. Working backward from the project's finish, we find the l.s.t. of the

several activities by assigning times to the nodes and arcs, according to the following rules:
1. Assign the target time, τ, to the finish node.
2. To each arc, assign the time of the node at which the arc ends, minus the length of the arc.
3. To each node, assign equal to the minimum of the times assigned to the arcs that begin at the node.

The times assigned to the nodes in this manner are the l.s.t.'s. Figure VII.2.3 shows the network of Example VII.2.1 with l.s.t.'s included.

FIGURE VII.2.3. Late start times (at nodes) for Example VII.2.1.

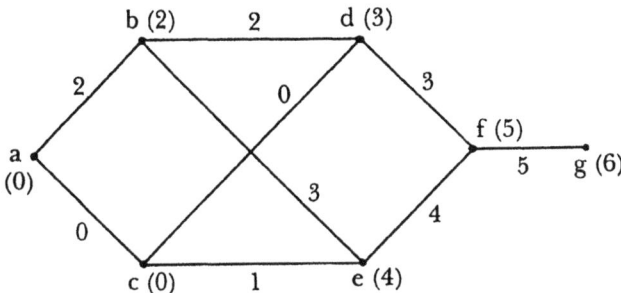

It is clear that the l.s.t. can never be smaller than the e.s.t. On the other hand, the e.s.t. might be smaller than the l.s.t. We shall call the difference between these two, the *slack time* for the particular activity.

7.2.1   Slack time = late start time – early start time

We define an activity as *critical* if its slack time is 0. In the example, the critical activities are a, c, d, f, and g. Of these, a and g ("start" and "finish") are obviously critical, as they are for all projects – though of course they are not really activities.

In Figure VII.2.4, the critical activities for the project are shown. The arcs joining these nodes are shown by the heavier lines. It may be seen that these arcs, each of which starts and ends at a critical activity, form a path through the network. This is known as the *critical path* of the project; it is not difficult to see that the length of the path, i.e. the sum of the lengths of the arcs that form this path, is precisely equal to τ.

FIGURE VII.2.4. Slack times for Example VII.2.1. The heavy line shows the critical path.

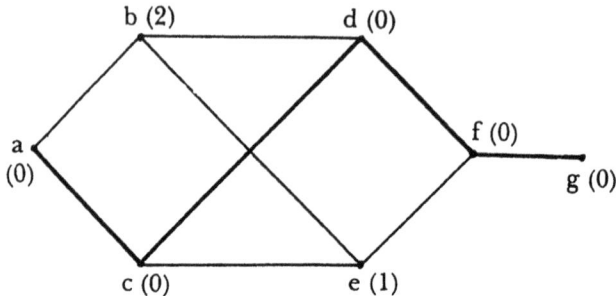

The critical activities are important in the following sense. Suppose it is decided that the project has to be completed in less than six days. This can only be done if some of the activities are carried out on a "crash" basis, say by paying substantial overtime to the workers or by paying a transport firm to ship things by air instead of by ground. Since, however, this can prove quite expensive, it follows that only a few, or, if possible, only one of the activities should be expedited in this manner. The problem is to determine which one. Clearly, it does no good to put activities b or e, the non-critical activities, on a crash basis. Not lying on the critical path, they cannot help to shorten it. Additional expenditures, if any, should only be made on the critical activities: *if the target time is to be decreased, it is always necessary to decrease the performance time for one or more of the critical activities.*

In the general case, similar considerations hold, the only difference being that there might be more than one critical path. To prove this, we use the following result:

VII.2.2. Lemma. In any project, a critical activity other than "start" or "finish" always has at least on critical activity among its immediate predecessors and at least one among its immediate successors. The activity "start" has at least one critical activity among its immediate successors, and "finish" has at least one critical activity among its immediate predecessors.

To prove the lemma, we note that an activity is critical if its e.s.t. and l.s.t. are equal. Suppose that activity j is critical. Then the e.s.t. of each successor of j is at least equal to the e.s.t. of j, plus j's time of performance. On the other hand, the l.s.t. of j is equal to the l.s.t. of some successor of j, which we shall call k, minus j's time of performance. Thus we have the relations:

$$\text{e.s.t. of } k \geq \text{e.s.t. of } j + \text{t.o.p. of } j$$
$$\text{l.s.t. of } j = \text{l.s.t. of } k - \text{t.o.p. of } j$$

which will give us

$$\text{e.s.t. of } k \geq \text{l.s.t. of } k$$

or

$$\text{slack time of } k \leq 0$$

Since the slack time is never negative, we conclude that k has slack time 0; i.e., k is critical. Thus j has a critical immediate successor. Similar considerations may be used to prove the rest of the lemma.

Let us see just what the lemma means. If we consider only those arcs of the network that begin and end at critical activities, the lemma states that each such arc, unless it ends at node "finish", must be followed by another; unless it starts at the node "start," it must be preceded by another. Each of these arcs is therefore on a path, made up entirely of critical arcs, going from "start" to finish." This is the *critical path* of the network. Every network has thus at least one critical path (but there may be several critical paths). The critical path is always the longest path through the network, and it is easy to see that, to decrease the target time of the project, it is necessary to decrease the time of performance of at least one activity on the critical path (or on each of the critical paths, if there are several such).

VII.2.3. Example. Consider the project of Example VII.2.1. Suppose that the time of performance for activity d (remove and paint door) is decreased from two days to one day. Since this is a critical activity, τ is decreased from six days to five days. The new project network, together with e.s.t.'s and l.s.t.'s, is shown in Figure VII.2.5. It may be seen that there are now two critical paths: (a, c, d, f, g) and (a, c, e, f, g). To decrease the target time further, we can decrease the time of performance of either activities c or f, which lie on both these paths. On the other hand, it will not be sufficient to decrease the time of performance of activity d, even though it is critical, because d lies on only one of the two critical paths. We could, however, decrease τ by decreasing the time of performance for both activities d and e, since this gives us one activity on each of the two critical paths.

VII.2.4. Example. A contractor makes plans for building a house as shown in Table VII.2.2.

FIGURE VII.2.5. Early and late start times for Example VII.2.3. The heavy lines show the critical paths.

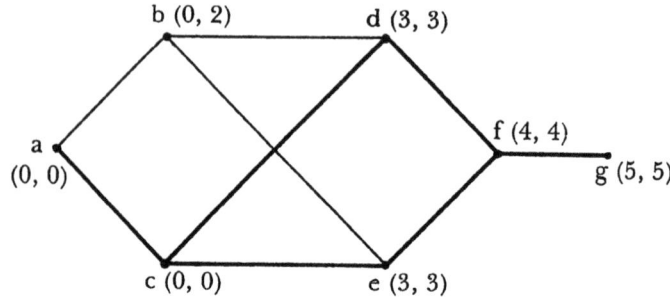

FIGURE VII.2.6. Early and late start times for Example VII.2.4. The critical path is shown.

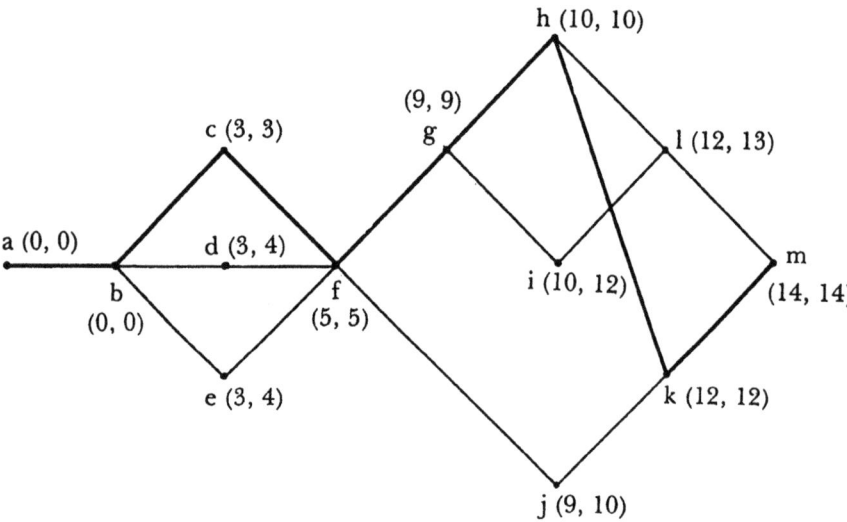

TABLE VII.2.2

Activity	Time (weeks)	Immediate predecessors
a. Start	0	None
b. Lay foundation	3	a
c. Construct basement	2	b
d. Underground wiring	1	b
e. Underground pipes	1	b
f. Construct main floor	4	c, d, e
g. Install fixtures	1	f
h. Connect plumbing	2	g
i. Connect wiring	1	g
j. Place roof	2	f
k. Roof drains	2	h, j
l. Paint house	1	h, i
m. Stop	0	k, l

Figure VII.2.6 shows the network for this project. The target time is 14 weeks, and, as may be seen, the critical path passes through nodes (a, b, c, f, g, h, k, m). The other nodes all have one or two weeks slack.

In practice, the times of completion of the several activities in a network are usually uncertain. In fact, there are always unforeseen delays connected with any project. A project manager will generally ask an expert for the estimated time of completion of each activity, and the expert will be the first to admit the existence of these uncertainties. When the probabilities of such delays are not negligible, it is sometimes necessary to ask the expert for a description of the probability distribution of each time of completion, which is then treated as a random variable. The means and variances of these random variables are then obtained. (It must, of course, be remembered that in many cases these are, at best, informed estimates, and at worst, wild guesses.) The critical path method is then used with the *means* of these times of completion as the lengths of the arcs.

If the non-critical activities have enough slack time, and if the *variances* of the times of completion are small enough, then the critical path obtained in this manner will, with high probability, be the critical path for the *actual* times of completion. (There is, unfortunately, always the probability that a non-critical activity will suffer an unexpected delay that will use up all of its slack time and more, but this has low probability.) Then the actual target time will be a random variable, equal to the sum of the times of completion of the activities on the critical path. In turn, these can generally be considered as independent random variables. The mean of their sum will be the sum of their means, while the variance of their sum will be the sum of their variances.

VII.2.5. Example. Consider once again the project of Example VII.2.1, but suppose a mechanic tells us that there is a "likely deviation" of 1 day in the time of completion of activity c (ordering parts) and of ¼ day in the times of completion of each of the other activities (excepting, of course, "start" and "finish"). We interpret this to mean that each time of completion is a random variable with mean equal to the value shown and a *standard deviation* of 1 day for activity c and ¼ day for the other activities.

With this interpretation, the critical path is unchanged; the target time will now be a random variable with mean equal to 6 days. To find its variance, we add the variances of activities c, d, and f, which are the squares of 1/4, 1, and 1/4 respectively. Thus,

$$\sigma^2 = 1/16 + 1 + 1/16 = 18/16$$

and so

$$\sigma = 1.06.$$

We conclude that τ is a random variable with mean 6 days and a standard deviation of slightly over 1 day.

PROBLEMS ON CRITICAL PATH ANALYSIS

1. The project of installing new light cables for an auditorium is described in the following table:

Activity	Time (days)	Immediate predecessors
a. Start	0	None
b. Obtain specifications	4	a
c. Make duplicates	3	b
d. Order cable	1	b
e. Order clamps	3	c
f. Order cutter	3	b
g. Thread cable	1	c, d
h. Cut cable	2	f, g
i. Install cable	3	e, h
j. Stop	0	i

(a) Find the e.s.t and l.s.t for each activity in this project.
(b) What is the target time?
(c) Find the critical path.

2. In preparing a shop for rental, a contractor must perform the following tasks:

Activity	Time (weeks)	Immediate predecessors
a. Start	0	None
b. Draw blueprints	2	a
c. Hire mechanics	1	a
d. Order parts	4	b
e. Duplicate blueprints	1	b
f. Adjust equipment	2	c, d, e
g. Hire subcontractors	3	e
h. Contract utilities	2	e
i. Install equipment	5	g, f
j. Landscape	2	c, g
k. Stop	0	h, i, j

(a) How long will it take to complete this project?
(b) Find the critical path.

3. In critical path analysis, the *free slack* of an activity is the amount of time that an activity can be delayed without delaying any of the other activities. The

free slack of *i* is equal to the minimum of the early start times of the successors of *i*, minus the early finish time of *i*. Find the free slack of the several activities in the project of problem 2 above.

3. The *independent slack* of activity *i* is the amount

$$A - B - C$$

where *A* is the minimum of the early start times of the successors of *i*, *B* is the time of performance of *i*, and *C* is the maximum of the late finish times of the predecessors of *i*. The independent slack is the amount of leeway available in scheduling activity *i*, without in any way affecting the scheduling of the other activities. Find the independent slack of the several activities involved in the project f Problem 2 above. (In general, very few activities will have any independent slack.)

3. THE SHORTEST PATH THROUGH A NETWORK

In the previous section, we introduced a network problem and saw that it reduces, in essence, to finding the longest path through a network. We are often faced, instead, with the problem of finding the *shortest* path between two nodes in a network. When we consider that the length of an arc can represent not only geographical distance, but also time required or costs, we see that this is a very common problem. We shall approach it through what is, essentially, a dynamic programming technique.

In general, the problem of finding the shortest path is not substantially different from that of finding the critical path. One principal difference is, perhaps, the fact that the new problem does not require the network to be oriented: there may be "two-way" arcs between the nodes. We shall assume throughout that all arcs are two-way. Practically, this may mean that the distance between two points is the same in either direction (which is a well-known fact) or that the transportation costs between two points are the same in either direction (which is not so certain). At any rate, the generalization to oriented graphs is easy and does not require different treatment.

The method we use for this problem depends on the observation that, if a network contains *n* nodes, then the shortest path through the network (between any two nodes) will require at most *n*-1 arcs (since more arcs would imply passing through the same node twice). The principle of optimality then states that, if a path is minimal, any sub-path will also be minimal.

Mathematically, we proceed as follows. From the graph, a table of distances d_{ij} is formed, in which d_{ij} is the length of the arc from i to j, if such an arc exists, and ∞, if there is no such arc. Let us now use the notation

$$F_m(i,j)$$

to denote the length of the shortest path from node i to node j, using at most m arcs. By the principle of optimality, it will follow that the functions F_m satisfy the equation

8.3.1 $$F_m(i,j) = \min_k \{F_{m-1}(i,k) + d_{kj}\}$$

We also have

8.3.2 $$F_1(i,j) = d_{ij}$$

Using this formulation, it is quite straightforward to use dynamic programming for the solution of these problems.

VII.3.1. Example. A motorist wishes to travel from one city to another by the shortest possible path. A road map shows the network of routes which is shown schematically in Figure VII.3.1. The motorist starts from city 1 and wishes to reach city 11.

FIGURE VII.3.1. Example VII.3.1.

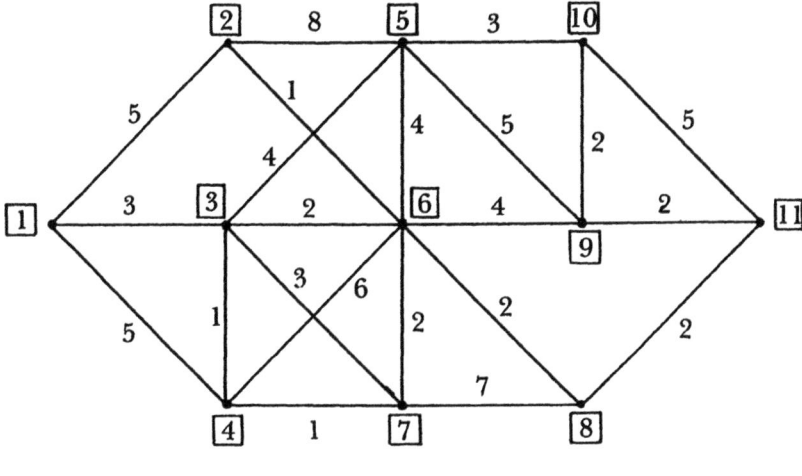

We tabulate the distances as follows:

TABLE VII.3.1

	1	2	3	4	5	6	7	8	9	10	11
1	0	5	3	5	--	--	--	--	--	--	--
2	5	0	--	--	8	1	--	--	--	--	--
3	3	--	0	1	4	2	3	--	--	--	--
4	5	--	1	0	--	6	1	--	--	--	--
5	--	8	4	--	0	4	--	--	5	3	--
6	--	1	2	6	4	0	2	2	4	--	--
7	--	--	3	1	--	2	0	7	--	--	--
8	--	--	--	--	--	2	7	0	--	--	2
9	--	--	--	--	5	4	--	--	0	2	2
10	--	--	--	--	3	--	--	--	2	0	5
11	--	--	--	--	--	--	--	2	2	5	0

Where the entries – mean that there is no direct-line connection between the corresponding cities. The quantity that we are looking for is, of course, $F_{10}(1,11)$. We shall compute the several functions $F_m(1,j)$ by equations (7.3.1) and (7.3.2).

The function $F_1(1,j)$ is given by (7.3.2) and be found in the table. We compute the functions $F_2(1,j)$ and $k_1^*(1,j)$ by

$$F_2(1,j) = \min_k \{F_1(1,k) + d_{kj}\}$$

and

$$F_2(1,j) = F_1(1, k_1^*) + F_1(k_1^*,j).$$

In other words, $k_{m-1}^*(1,j)$ is defined by saying that to go from 1 to j in m steps, the $(m-1)$th step should go through node k_{m-1}^*. We have, then,

TABLE VII.3.2

j	$F_2(1,j)$	$k_1^*(1,j)$
1	0	1
2	5	2
3	3	3
4	4	3
5	7	3
6	5	3
7	6	3

This says, for instance, that the shortest two-step path from node 1 to node 6 is 5 miles long, and passes through node 3. The two-step distances are not given for the other nodes, because these cannot be reached in two steps.

Proceeding in the usual manner, we construct a table of the functions $F_m(1,j)$ and $k_{m-1}*(1,j)$, given in Table VII.3.3.

TABLE VII.3.3

j	$F_2(1,j)$	$F_3(1,j)$	$F_4(1,j)$	$k_1*(1,j)$	$k_2*(1,j)$	$k_3*(1,j)$
1	0	0	0	1	1	1
2	5	5	5	2	2	2
3	3	3	3	3	3	3
4	4	4	4	3	3	3
5	7	7	7	3	3	3
6	5	5	5	3	3	3
7	6	5	5	3	4	4
8	--	7	7	--	6	6
9	--	9	9	--	6	6
10	--	10	10	--	5	5
11	--	--	9	--	--	8

Values F_m can be computed for $m = 5, 6$, and so on, but these are not necessary. In fact, if we compute them, we shall see that the distances are not decreased: the minimal path between node 1 and any other node in the network always takes 4 steps or less. The solution is obtained, as usual, by working backward. We have $k_3*(1,11) = 8$, $k_{12}*(1,8) = 6$, and $k_1*(1,6) = 3$. Thus the shortest path from node 1 to node 11 is 9 miles long, passing through 3, 6, and 8 on the way. Figure VII.3.2 shows the minimal path in the network.

As can be seen from the4 example, this is strictly a dynamic programming technique. Several types of modification are possible. We may, for example, compute the distances $F_m(i,11)$ rather than $F_m(1,j)$. This would tell us how to proceed so as to reach destination node 11 from any starting point. The technique we used tells us, instead, how to reach any destination from the fixed starting node 1. Depending on the actual purpose of the decision-maker, one technique may be more useful than the other. There is not, generally, any difficulty in effecting a variation.

FIGURE VII.3.2. Minimal path for Example VII.3.1.

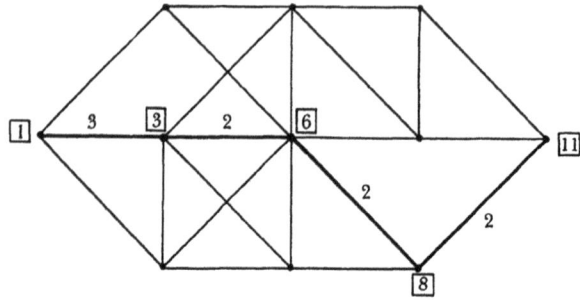

PROBLEMS ON SHORTEST PATHS THROUGH NETWORKS

1. A traveler wishes to fly from Central City to Capital City, some 1700 miles away. Unfortunately, he has only a very small plane, with a flying range of 800 miles. He must, therefore, make several fueling stops along the way. There are twelve intermediate cities with airports. The distances among these cities, in tens of miles, (with 0 for Central City and 13 for Capital City) are given in the table. (A blank space means that the distance is greater than 800 miles and thus unmanageable for the plane.)

	0	1	2	3	4	5	6	7	8	9	10	11	12	13
0	0	51	58	60	61	--	--	--	--	--	--	--	--	--
1	51	0	24	16	21	56	55	52	63	61	--	--	--	--
2	58	24	0	17	20	52	56	67	61	54	--	--	--	--
3	60	16	17	0	17	51	58	55	53	50	--	--	--	--
4	61	21	20	17	0	51	52	60	58	48	--	--	--	--
5	--	56	52	51	51	0	22	24	18	21	52	58	60	--
6	--	55	56	58	52	22	0	22	19	30	56	52	51	--
7	--	52	67	55	60	24	22	0	22	22	59	62	57	--
8	--	63	61	53	58	18	19	22	0	27	55	56	52	--
9	--	61	54	50	48	21	30	22	27	0	52	54	57	--
10	--	--	--	--	--	52	56	59	55	52	0	32	21	43
11	--	--	--	--	--	58	52	62	56	54	32	0	27	44
12	--	--	--	--	--	60	51	57	52	57	21	27	0	51
13	--	--	--	--	--	--	--	--	--	--	43	44	51	0

What is the shortest route available to the traveler?

2. Find the shortest path through the graph in figure VII.3.3, from Node 0 to node 11.

FIGURE VII.3.3.

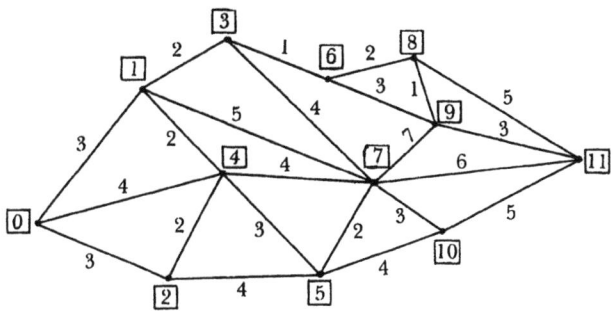

4. MINIMAL SPANNING TREES

We consider a slightly different problem, that of finding *a minimal spanning tree* in a network.

To understand the problem, let us suppose that a utility company must service a set of customers. It must then construct lines that connect it with each of the customers, so as to deliver the product. It is not necessary to draw independent lines from the utility to each customer; a single line, passing by each customer, will do; so will several lines, each servicing a small number of customers. In general, it is necessary only that the lines all be connected and reach all of the customers. In other words, the customers must be the nodes of a connected graph; the utility lines will be the arcs of this graph. The company will naturally try to build the lines as economically as possible, and so we look for the minimal cost graph that can be built subject to these restrictions. In general, we will start with a graph showing all possible distances between nodes (i.e., between pairs of customers or between the company and a customer); the problem is to choose the arcs that will give a minimal total distance. It is not too difficult to see that these lines will never form a loop; we talk, then, about the *minimal tree* that *spans* the network (i.e., passes through all the nodes).

The method we use for this problem is an extremely simple one, known as *Kruskal's algorithm*. This algorithm tells us to build the tree by adding arcs, one at a time, according to the rule: *of all the arcs that are not yet part of the tree, and* that do not form a loop when added to those already in the tree, choose the shortest one. (In case of ties, any one of those tying for shortest may be added to the tree.)

FIGURE VII.4.1. A spanning tree.

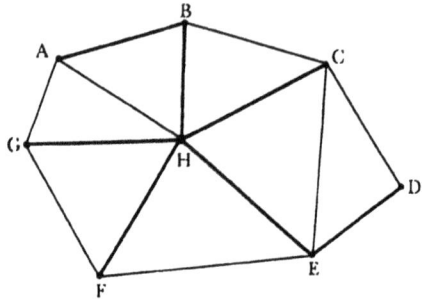

FIGURE VII.4.2. This tree does not span the network, node C is isolated.

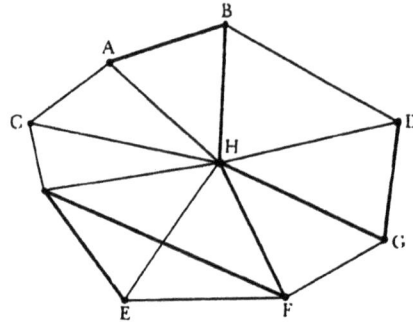

To prove that Kruskal's algorithm does indeed give us the minimal spanning tree, we make a non-degeneracy assumption: no two arcs have equal lengths. If, in fact, this is not so, we will make use of a perturbation technique: when two or more arcs have equal lengths, these lengths are altered by small amounts ε, 2ε, 3ε, and soon. If ε is small enough, the minimal tree for the perturbed network will still be minimal for the original, unperturbed problem.

VII.4.1. Proof of Kruskal's Algorithm. To see that the algorithm will give us a connected graph, note that, so long as the graph is disconnected, there are still arcs that can be added without closing a loop. Thus, the algorithm will stop only when a connected graph is obtained. Clearly, by the rule of construction, this graph can have no loops and so it must be a tree, ant it follows that it will be a spanning tree. Let us call this tree H. (We think of H as a collection of arcs and nodes.)

Suppose, now, that H is not the minimal spanning tree. This means that there is a minimal one, W, which is different from H. The two trees, H and W, have the same number of arcs, namely, one fewer than the number of nodes. Since they do not have the same arcs (as this would make them identical) it follows that H has at least one arc that is not in W. Let (i,j) be the shortest arc that is in H but not in W. Then $W \cup \{(i,j)\}$ is a graph with n nodes and n arcs, and therefore contains a loop. In this loop, there is an arc (k,l) which belongs to W but not to H (since H does not contain any loops). Consider then the graph

$$F = W \cup \{(i,j)\} - \{(k,l)\}$$

The graph F is connected (since removing an arc from a loop does not disconnect a graph) and has n nodes and n-1 arcs. Therefore F is a spanning tree. Now the length of F is equal to the length of W, plus the length of (i,j), minus the length of (k,l).

Suppose, now, that (k,l) were shorter than (i,j). We know all the arcs in H which are smaller than (i,j) (i.e., those which were chosen before (i,j) was chosen) are also in W. Thus, (k,l) cannot form a loop when added to these arcs. But this would mean that (i,j) was added to the graph H against the algorithm's rules (i.e., it was not the shortest available), and we conclude that (k,l) is not shorter than (i,j). By the non-degeneracy assumption, (i,j) is shorter than (k,l). This means that F is shorter than W, contradicting the minimality of W. The contradiction tells us that H is the desired minimal tree.

We conclude this section with an example.

VII.4.2. Example. Find the minimal spanning trees for the networks shown in Figures VII.4.3 and VII.4.4.

For Figure VII.4.3, we successively take the following arcs: *EH, FG, BE, DG, EG, AB, AC*. For Figure VII.4.4, we take *AB, FG, BH, BF, CG, CD, EF*. Figures VII.4.5 and VII.4.6 show the minimal trees on those networks.

FIGURE VII.4.3. Example VII.4.2.

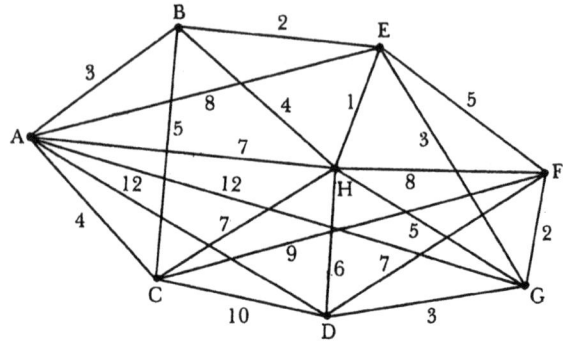

FIGURE VII.4.4. Example VII.4.2.

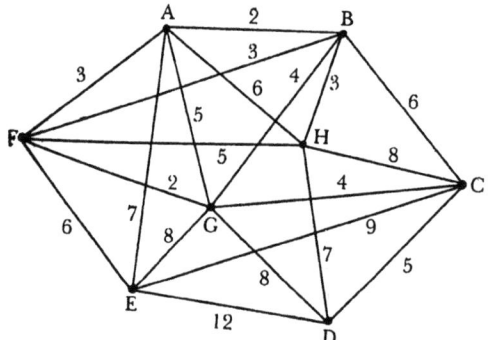

FIGURE VII.4.5. Solution to Example VII.4.2: minimal spanning tree for Figure VII.4.3.

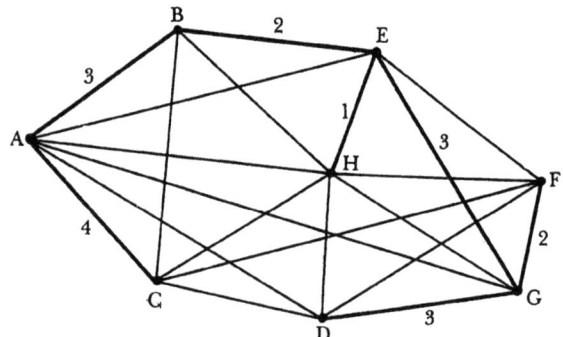

FIGURE VII.4.6. Solution to Example VII.4.2: minimal spanning tree for Figure VII.4.4.

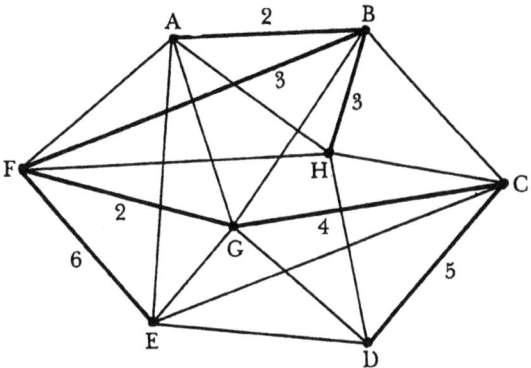

PROBLEMS ON MINIMAL SPANNING TREES

1. A company wishes to connect its 10 branches by means of an intranet system. The distances between branches are given in the table:

	1	2	3	4	5	6	7	8	9	10
1	0	23	42	31	27	38	29	33	41	25
2	23	0	29	36	25	34	44	27	31	29
3	42	29	0	31	38	27	25	41	43	28
4	31	35	31	0	26	36	33	45	37	26
5	27	25	38	26	0	31	37	32	41	28
6	38	34	27	36	31	0	25	24	38	33
7	29	44	25	33	37	25	0	31	36	29
8	33	27	41	45	32	24	31	0	41	27
9	41	31	43	37	41	38	36	41	0	32
10	25	29	28	26	28	33	29	27	32	0

Assume that the cost of connecting two branches directly is proportional to the distance between them, and that messages from one branch to another can be relayed with no loss in efficiency. What is the most efficient way to connect the system?

2. Find the minimal spanning tree in Figures VII.4.7, VII.4.8, VII.4.9, and VII.4.10.

FIGURE VII.4.7

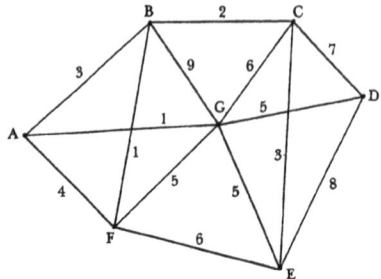

FIGURE VII.4.8

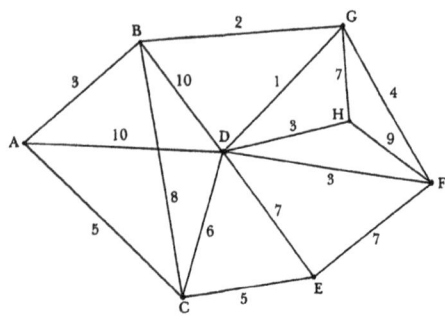

FIGURE VII.4.9

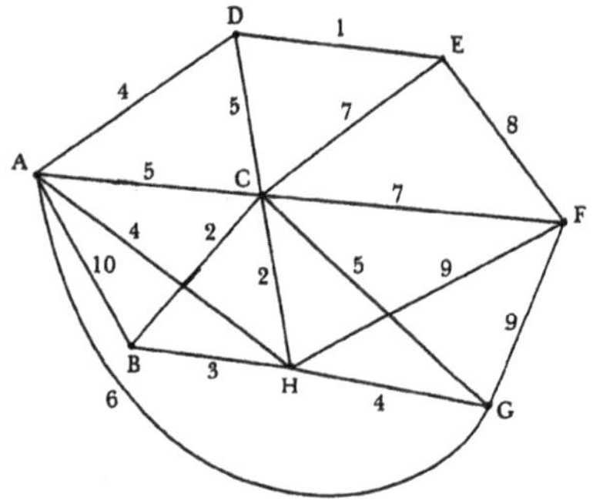

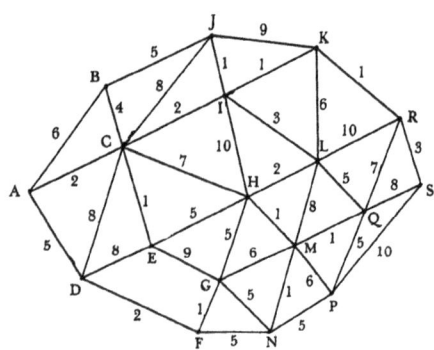

FIGURE VII.4.10

5. THE MAXIMAL FLOW IN A NETWORK

In this section, we consider a considerably more complicated problem, that of finding the maximal flow through a network.

In essence, we are given a network with two distinguished nodes, called a *source* and a *sink*. To each arc is assigned a number representing not a cost or a distance, but a *capacity*. We assume that some material is to be shipped from the source to the sink over the arcs of the network; the problem is to find the maximal amount that can be shipped, subject to the constraint that the amount shipped over any arc can be no greater than the capacity. We should note here that this is assumed to be a directed graph: the capacity from i to j is not necessarily the same as from j to i -- in fact there may be no capacity at all in one direction of an arc.

Mathematically, we can formulate this problem as a linear program. The source and the sink will be called nodes 0 and n, respectively; the low (i.e., the amount shipped) from node i to node j will be represented by x_{ij}, and must be no greater than the arc capacity, c_{ij}. The problem is then to

Maximize

7.5.1 $$\sum_{j=1}^{n} x_{0j} - \sum_{i=1}^{n} x_{i0} = w$$

Subject to

7.5.2 $$\sum_{j=1}^{n} x_{ij} = \sum_{j=1}^{n} x_{ji} \quad \text{for each } i$$

and

7.5.3 $$x_{ij} \leq c_{ij} \quad \text{for each } i, j$$

7.5.4 $$x_{ij} \geq 0 \quad \text{for each } i, j$$

Note that constraint (7.5.2) means that, at each node other than the source or sink, the amount flowing into the node must equal the amount flowing out of the node; (7.5.3) means that the amount in any arc must be no greater than the arc capacity, and (7.5.4), that only positive flows are considered. The objective function (7.5.1) is the net amount leaving the source, which because of the conservation constraints (7.5.2), is also the net amount flowing into the sink.

Generally speaking, if the network is not too complex, it is possible to obtain the maximal flow by inspection. In the network of Figure VII.5.1(a), the maximal flow is easily seen to be 6 units, as shown in Figure VII.5.1(b). It is clear that this flow is maximal since the arcs leading from the source have no more capacity.

Figure VII.5.2 shows a slightly more complicated network, but even so, it is not difficult to obtain the maximal flow. The principal problem lies in getting 5 units to node 4. A reasonable method would be to take the maximal flow along the top route; this gives us a starting flow of 3 units. These capacities can be subtracted from the network, leaving a smaller network, as shown in Figure VII.5.3. Again we take the top route of this new network, which has a capacity of 1 units, and subtract these capacities to give the network shown in Figure VII.5.4. We proceed to do this several times, until finally the maximal flow, shown in Figure VII.5.5, is obtained.

FIGURE VII.5.1. Capacities (a) and maximal flow (b) in a network.

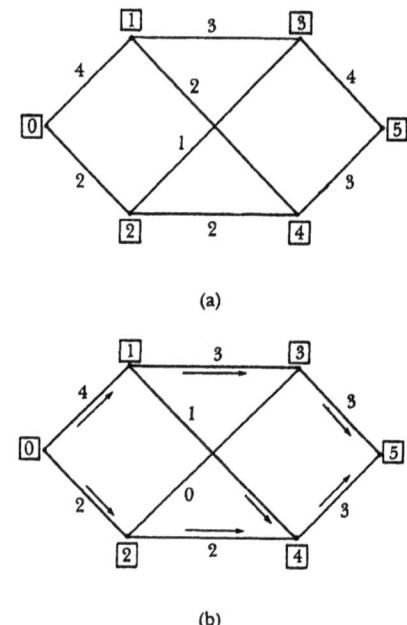

FIGURE VII.5.2. A flow network with capacities.

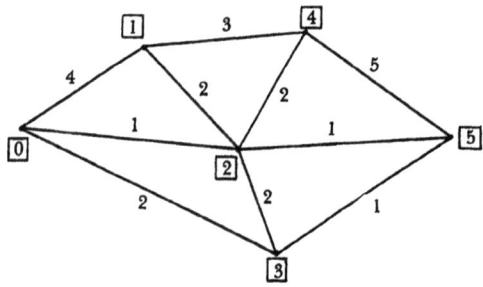

FIGURE VII.5.3. We take the maximal flow along the "uppermost" route.

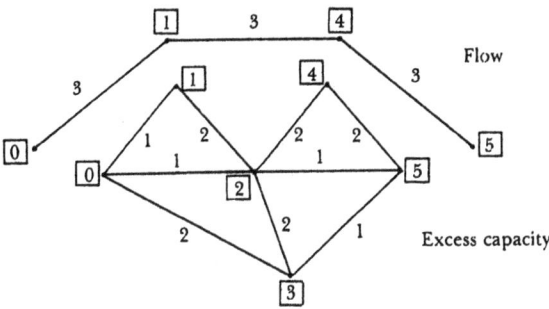

FIGURE VII.5.4. A somewhat lower route is taken.

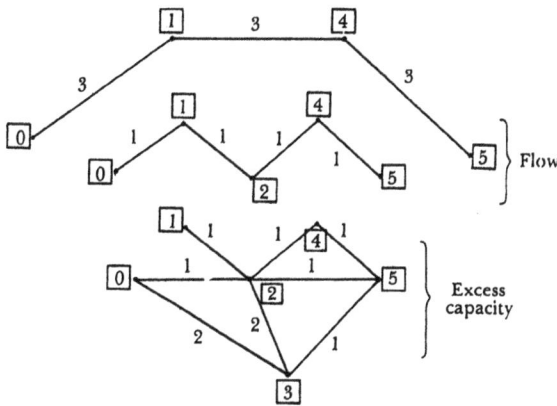

The method used for this example is quite straightforward and intuitive. For simple networks it is probably the easiest to use. Unfortunately, with large and complicated networks, it becomes almost impossible to keep track of all these excess capacities without some bookkeeping techniques. This is especially so if the network is non-planar; i.e. if some of the arcs cross (on paper) without actually meeting at a node.

This being a linear programming problem, it can be solved by the simplex algorithm. Once again, however, we find that the are shorter methods, thanks to the special form of the problem. The particular method that we shall use, developed by Ford and Fulkerson, is known as the labeling technique.

We start with two definitions.

FIGURE VII.5.5. Maximal flow for Figure VII.5.2.

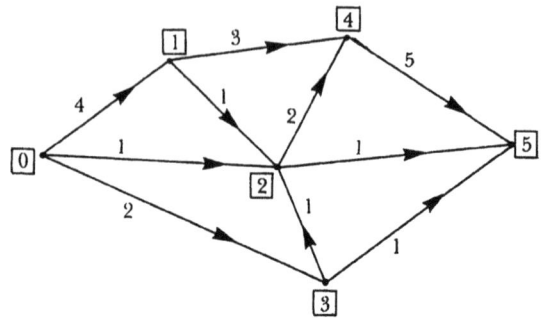

VII.5.1. Definition. We shall say that the arc (i,j) is *saturated* if $x_{ij} = c_{ij}$. It is *empty* if $x_{ij} = 0$.

VII.5.2. Definition. We shall say that the arc (i,j) has *positive excess capacity* if either (a) (i,j) is unsaturated or (b) the opposite arc (j,i) is non-empty.
In case (a), the excess capacity is

7.5.5 $$g_{ij} = c_{ij} - x_{ij}$$

and, in case (b),

7.5.6 $$g_{ij} = x_{ij}$$

Essentially, the excess capacity of a given arc is the amount that it can carry, over and above what it is already carrying. The flow along that arc may be increased by the amount of excess capacity. In case (b), in which the flow is in the opposite direction, this reverse flow (from j to i) may be <u>decreased</u> by the amount of excess capacity, which from our point of view has the same effect as increasing the flow from i to j by that amount.

An alternative definition of excess capacity could be

7.5.7 $$g_{ij} = c_{ij} - x_{ij} + x_{ji}$$

We now describe the labeling procedure, giving it as a set of rules. It is assumed that a feasible flow has been given. Each iteration starts with a feasible flow; the purpose of the iteration is to obtain a new (greater) feasible flow.

VII.5.3. Rules for the Labeling Procedure.

1. Start at the source (node 0). If any of the arcs $(0,j)$ has positive excess capacity, assign the label (t_j, k_j), where

$$t_j = g_{0j}$$
$$k_j = 0$$

to the corresponding vertex j.

2. (General Step) From among the labeled nodes, let i be the smallest index which has not yet been treated (in the iteration). If there are any arcs (i,j) with positive excess capacity, starting at this node, and such that their terminal nodes j *have not yet been labeled in this iteration*, then assign the label (t_j, k_j), where

7.5.8 $$t_j = g_{ij}$$

7.5.9 $$k_j = i$$

to the corresponding vertex j.

3. Repeat step 2 as many times as possible.

The rules VII.5.3 tell us how to carry out each iteration of the labeling procedure. In effect, if node j has been labeled (t_j, k_j), that means that it is possible to send additional flow from the source to j. This additional flow will follow a sequence of arcs that passes through k_j immediately before j.

If we follow the rules VII.5.3, the iteration will stop when either (a) the sink, node n, has been labeled, or (b), it is impossible to label any more nodes, and n remains unlabeled. This will occur in at most n steps, since each step in the iteration treats one of the n nodes (not the sink) in the network, and no node will be treated more than once in the same iteration. We shall show that, in case (a), the flow can be increases. In case (b), the given flow is maximal.

Suppose, indeed, that case (a) holds: the sink has been labeled. In this case, some additional flow can be sent from the source to the sink, along a path of arcs on the graph. The labels tell us the path to be followed. In fact, $h = k_n$ will be the next to last node on this path. Then k_h will be the node before h, and so on. The path of the new flow, then, will be composed of the arcs (let us say) $(0, j_1), (j_1, j_2), \ldots, (j_m, n)$. The flow along each of these arcs can be increased by as much as the minimum of the excess capacities on this path. These excess capacities are given by the labels t_j.

Thus, the smallest of the t_j on the path is the amount of additional flow that can be sent.

Now, the network is essentially unchanged if the capacities are all multiplied by the same positive constant. Then, if the capacities are all rational numbers, we can multiply all of them by their common denominator so that all of them become integers. Thus, it is not a great restriction to assume that all capacities are integers.

Assuming this, the flows obtained in this fashion will always be integers, as will the excess capacities. It will follow that, at each iteration of the labeling procedure, the flow will increase by at least one unit. Since the flow is clearly bounded above (i.e., the linear program (7.5.1) to (7.5.4) is bounded) it follows that the rules VII.5.3 can only be followed a finite number of times; eventually a flow will be obtained for which the sink cannot be labeled.

We next show that, if the sink cannot be labeled, then the flow in the network is maximal. To do this, we define a *cut* in the network.

VII.5.4. Definition. A cut in the network is a partition of the set N of all nodes in the network into two sets, (L, U) such that

$$0 \in L$$
$$n \in U$$
$$U \cap L = \emptyset$$
$$U \cup L = N$$

VII.5.5. Definition. Let (L, U) be a cut in a network. Then the *value* of the cut is the quantity

7.5.10
$$v(L, U) = \sum_{i \in L} \sum_{j \in U} c_{ij}$$

In essence, the value of a cut is the total capacity leading from set L into set U. Intuitively, it is clear that any amount shipped from the source to the sink must pass through one of those arcs, and it follows that the flow in a network can never be greater than the value of a cut. Mathematically, we prove this by defining the *flow across a cut* (L, U) as

7.5.11
$$f(L, U) = \sum_{\substack{i \in L \\ j \in U}} x_{ij} - \sum_{\substack{i \in L \\ j \in U}} x_{ji}$$

Then, by constraints (7.5.3) and (7.5.4), it is easy to see that $f(L,U) \le v(L,U)$. By constraints (7.5.2), we find that $w = f(L,U)$ for any cut (L,U). Thus,

$$w \le v(L,U)$$

Suppose, then, that we have a flow (x_{ij}), such that the sink cannot be labeled. Let L be the set of all nodes which can be labeled (including the source, which we think of as being labeled), and let U be the set of unlabeled nodes. Then (L,U) is a cut in the network. Now the rules VII.5.3 for the labeling procedure are such that, if i is labeled, and (i,j) has excess capacity, then j is also labeled. It follows that the arcs from L to U all have zero excess capacity. In other words, all arcs from L to U are saturated, while all arcs from U to L are empty. Thus,

$$\sum_{\substack{i \in L \\ j \in U}} x_{ij} = \sum_{\substack{i \in L \\ j \in U}} c_{ij}$$

and

$$\sum_{\substack{i \in L \\ j \in U}} x_{ij} = 0$$

so that

$$f(L,U) = v(L,U)$$

But this means that $w = v(L,U)$: the flow in the network is equal to the value of the cut (L,U). Since the flow can never exceed the value of a cut, we conclude that w is the maximal flow in the network.

We note that, in passing, we have proved an important theorem of network theory:

VII.5.6. Theorem (The Max-Flow Min-Cut Theorem). In any network, the maximum possible flow is equal to the value of the minimum cut.

It should be pointed out that VIII.5.6 is a special case of the duality theorem for linear programs. We illustrate the labeling technique by an example.

VII.5.7. Example. Find the maximum flow in the following network. In any arc, the flow is to be in the direction of the arrows only.

FIGURE VII.5.6. Capacities for Example VII.5.7. The flow (currently 0) can be increased along the route marked by the heavy line.

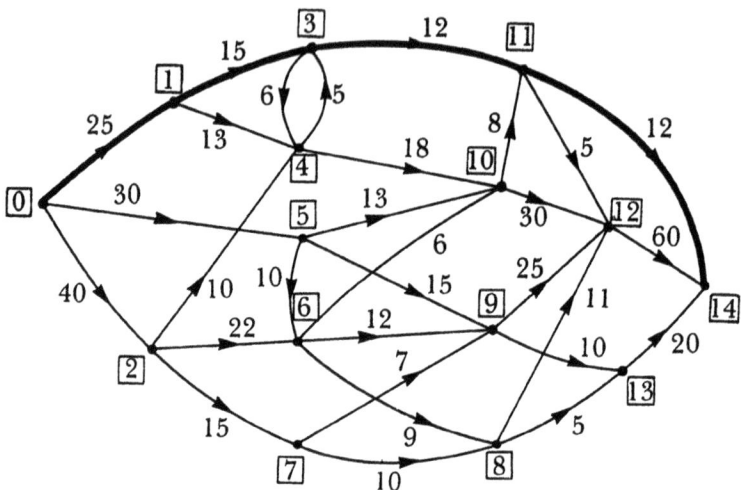

We start with the null flow, i.e., set $x_{ij} = 0$ for all i and j. The excess capacities g_{ij} are then equal to the capacities c_{ij}. We see that the source can be joined to nodes 1, 2, and 5 by arcs with positive excess capacity 25, 40, and 30 respectively, which allows to label these three nodes, with

$$\begin{array}{lll} t_1 = 25 & t_2 = 40 & t_5 = 30 \\ k_1 = 0 & k_2 = 0 & k_5 = 0 \end{array}$$

Now node 1 has been labeled, so we treat it next. It is joined to nodes 3 and 4 by arcs with excess capacities 15 and 13, so

$$\begin{array}{ll} t_3 = 15 & t_4 = 13 \\ k_3 = 1 & k_4 = 1 \end{array}$$

We treat next node 2. This can be joined to nodes 4, 6, and 7. Node 4 has already been labeled, so we set

$$t_6 = 22 \quad t_7 = 15$$
$$k_6 = 2 \quad k_7 = 2.$$

Next, node 3 can be joined to node 4, which is already labeled, and to node 11. This gives us

$$t_{11} = 12$$
$$k_{11} = 3$$

We proceed in this manner, treating the labeled nodes one at a time, until finally we obtain the following labels:

TABLE VII.5.1

j	1	2	3	4	5	6	7	8	9	10	11	12	13	14
t_j	25	40	15	13	30	22	15	10	15	18	12	11	5	12
k_j	0	0	1	1	0	2	2	7	5	4	3	8	8	11

We see here that the sink (node 14) has been labeled. To find the path from source to sink, we take

$$k_{14} = 11 \quad k_{11} = 3 \quad k_3 = 1 \quad k_1 = 0$$

which, taken backwards, gives us the path 0-1-3-11-14. The value of the flow will be the minimum of

$$t_{14} = 12 \quad t_{11} = 12 \quad t_3 = 15 \quad t_1 = 25$$

or 12. We add this flow to what is already in the network (0, in this case), and obtain the flow shown in Figure VII.5.7. At each arc, the first number represents the capacity; the second number, the flow.

Treating Figure VII.5.7 just as we treated VII.5.6, we eventually obtain

TABLE VII.5.2

j	1	2	3	4	5	6	7	8	9	10	11	12	13	14
t_j	13	40	5	13	30	22	15	9	15	18	8	11	5	60
k_j	0	0	4	1	0	2	2	6	5	4	10	8	8	12

FIGURE VII.5.7. Capacities (in large type) and flows (in smaller type). W = 12.

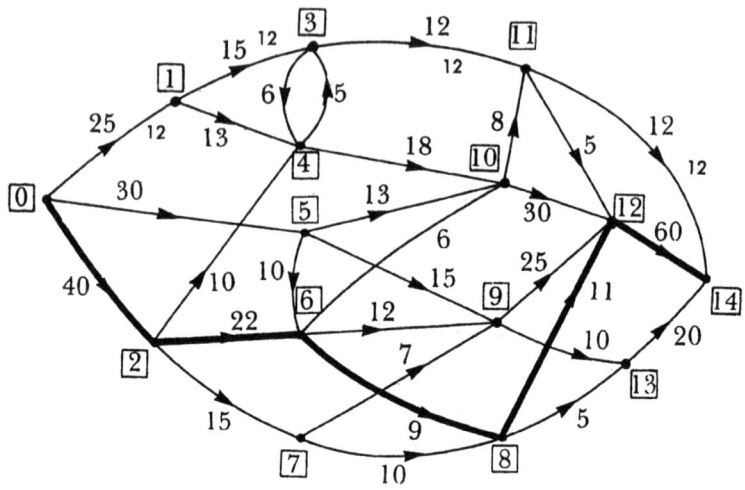

Once again, we find that the sink is labeled; we have

$$k_{14} = 12 \quad k_{12} = 8 \quad k_8 = 6 \quad k_6 = 2 \quad k_2 = 0$$

and

$$t_{14} = 60 \quad t_{12} = 11 \quad t_8 = 9 \quad t_6 = 22 \quad t_2 = 40$$

which means that 9 units may be sent along the path 0-2-6-8-12-14. This gives us Figure VII.5.8.

For Figure VII.5.8, we have

TABLE VII.5.3

j	1	2	3	4	5	6	7	8	9	10	11	12	13	14
t_j	13	31	3	13	30	13	15	10	15	18	8	2	5	51
k_j	0	0	1	1	0	2	2	7	5	4	10	8	8	12

We have, then,

$$k_{14} = 12 \quad k_{12} = 8 \quad k_8 = 7 \quad k_7 = 2 \quad k_2 = 0$$

and

$$t_{14} = 51 \quad t_{12} = 2 \quad t_8 = 10 \quad t_7 = 15 \quad t_2 = 31$$

FIGURE VII.5.8. W = 21.

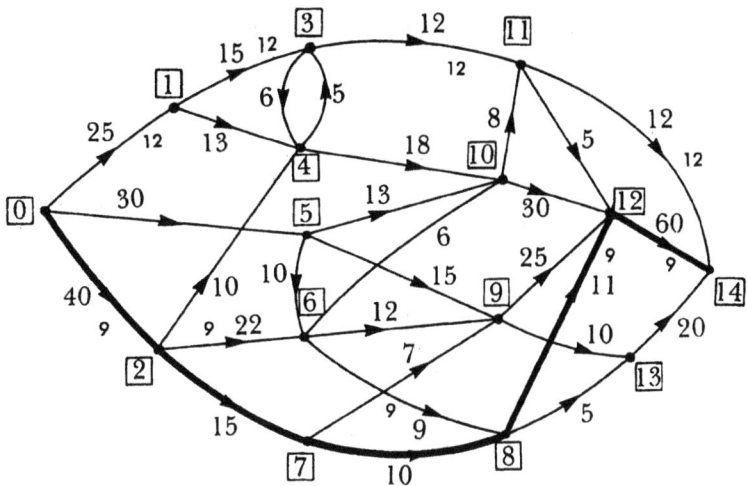

so that 2 units may be sent along the path 0-2-7-8-12-14. This gives us Figure VII.5.9 and Table VII.5.4.

FIGURE VII.5.9. W = 23.

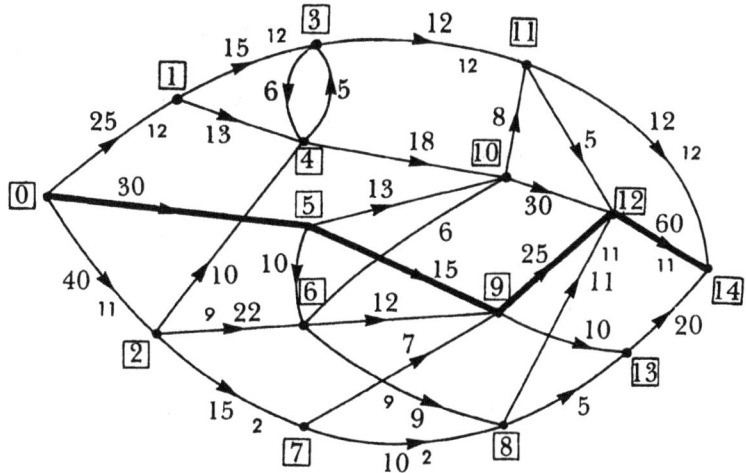

TABLE VII.5.4

j	1	2	3	4	5	6	7	8	9	10	11	12	13	14
t_j	13	29	3	13	30	13	13	8	15	18	8	25	5	49
k_j	0	0	1	1	0	2	2	7	5	4	11	9	8	12

This gives us

and

$$k_{14} = 12 \quad k_{12} = 9 \quad k_9 = 5 \quad k_5 = 0$$

$$t_{14} = 49 \quad t_{12} = 25 \quad t_9 = 15 \quad t_5 = 30$$

so that 15 more units can be sent along 0-5-9-12-14. This gives Figure VII.5.10 and Table VII.5.5.

FIGURE VII.5.10. W = 38.

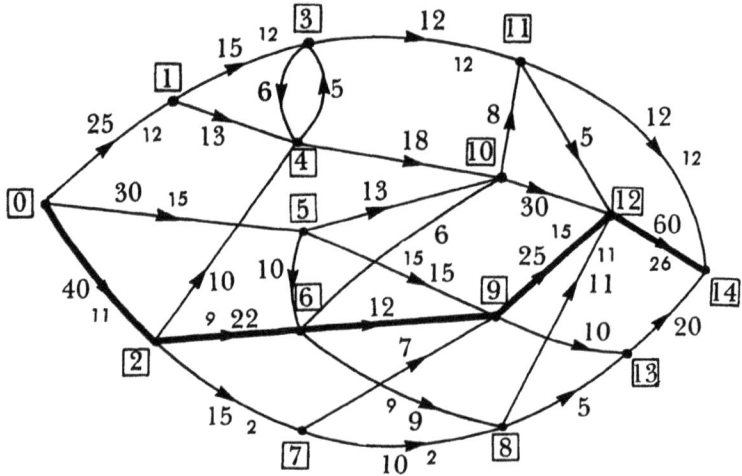

TABLE VII.5.5

j	1	2	3	4	5	6	7	8	9	10	11	12	13	14
t_j	13	29	3	13	15	13	13	8	12	18	8	10	5	34
k_j	0	0	1	1	0	2	2	7	6	4	10	9	8	12

We see here that 10 more units can be sent along 0-2-6-9-12-14. This gives Figure VII.5.11 and Table VII.5.6.

FIGURE VII.5.11. W = 48.

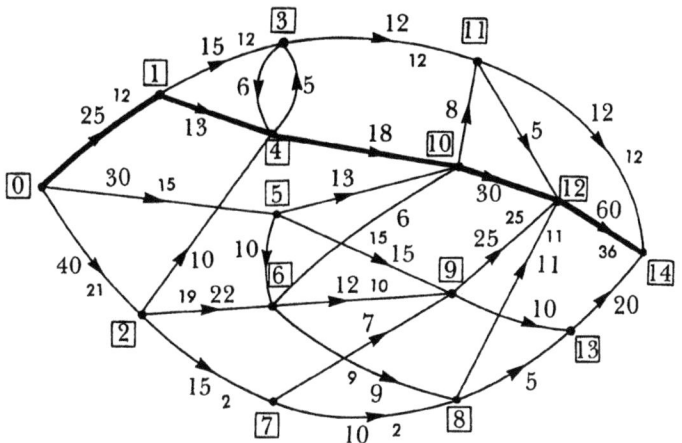

TABLE VII.5.6

j	1	2	3	4	5	6	7	8	9	10	11	12	13	14
t_j	13	19	3	13	15	3	13	8	2	18	8	30	5	24
k_j	0	0	1	1	0	2	2	7	6	4	10	10	8	12

Now 13 units can be sent along 0-1-4-10-12-14. This gives Figure VII.5.12 and Table VII.5.7.

FIGURE VII.5.12. W = 61.

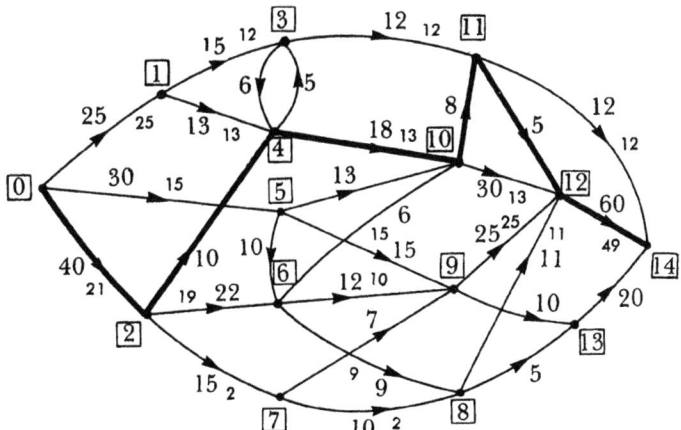

TABLE VII.5.7

j	1	2	3	4	5	6	7	8	9	10	11	12	13	14
t_j	13	19	5	10	15	3	13	8	2	5	8	5	5	11
k_j	4	0	4	2	0	2	2	7	6	4	10	11	8	12

We continue in this manner, obtaining Figures VII.5.12 through VII.5.18 and the corresponding tables VII.5.8 through VI.5.12.

FIGURE VII.5.13. W = 66.

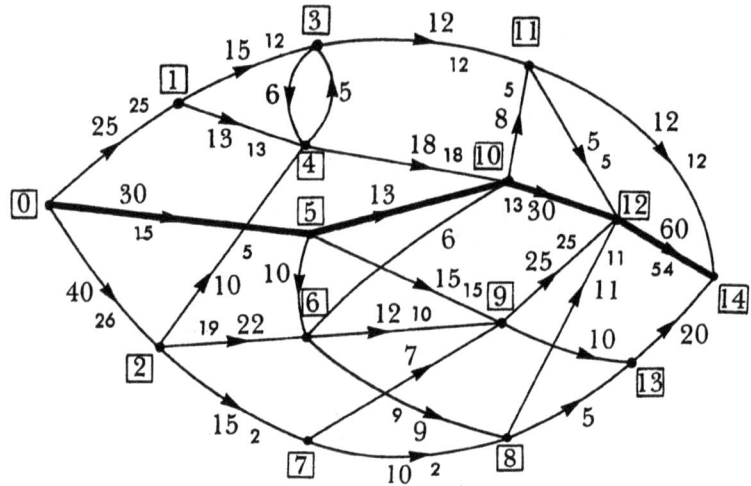

TABLE VII.5.8

j	1	2	3	4	5	6	7	8	9	10	11	12	13	14
t_j	13	14	5	5	15	3	13	8	2	13	3	17	5	6
k_j	4	0	4	2	0	2	2	7	6	5	10	10	8	12

FIGURE VII.5.14. W = 72.

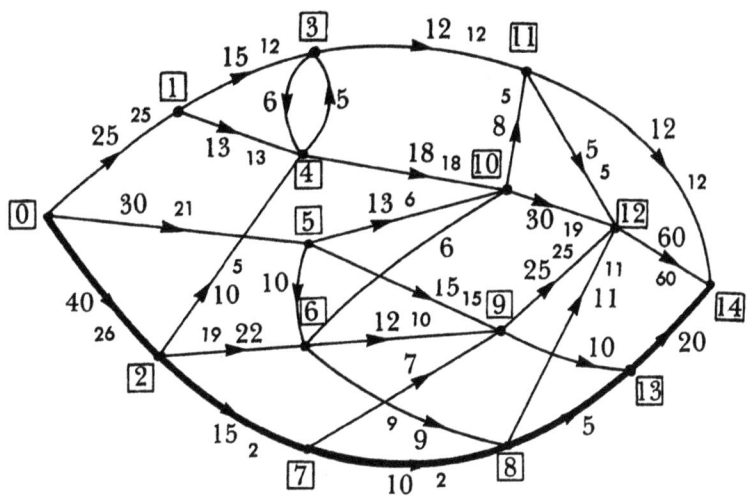

TABLE VII.5.9

j	1	2	3	4	5	6	7	8	9	10	11	12	13	14
t_j	13	14	5	5	9	3	13	8	2	7	3	11	5	20
k_j	4	0	4	2	0	2	2	7	6	5	10	10	8	13

FIGURE VII.5.15. W = 77.

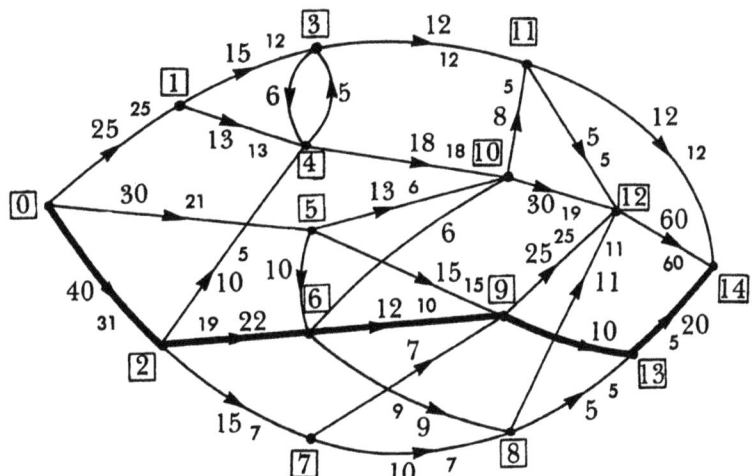

TABLE VII.5.10

j	1	2	3	4	5	6	7	8	9	10	11	12	13	14
t_j	13	9	5	5	9	3	8	3	2	7	3	11	10	15
k_j	4	0	4	2	0	2	2	7	6	5	10	10	9	13

FIGURE VII.5.16. W = 79.

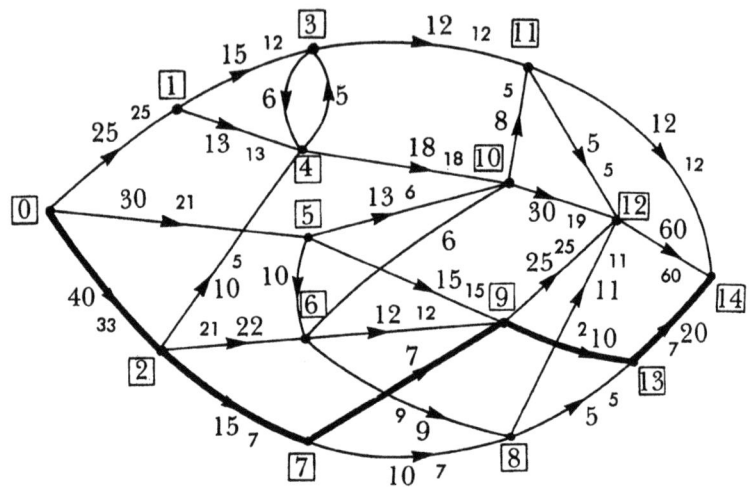

TABLE VII.5.11

j	1	2	3	4	5	6	7	8	9	10	11	12	13	14
t_j	13	7	5	5	9	1	8	3	7	7	3	11	8	13
k_j	4	0	4	2	0	2	2	7	7	5	10	10	9	13

FIGURE VII.5.17. W = 86.

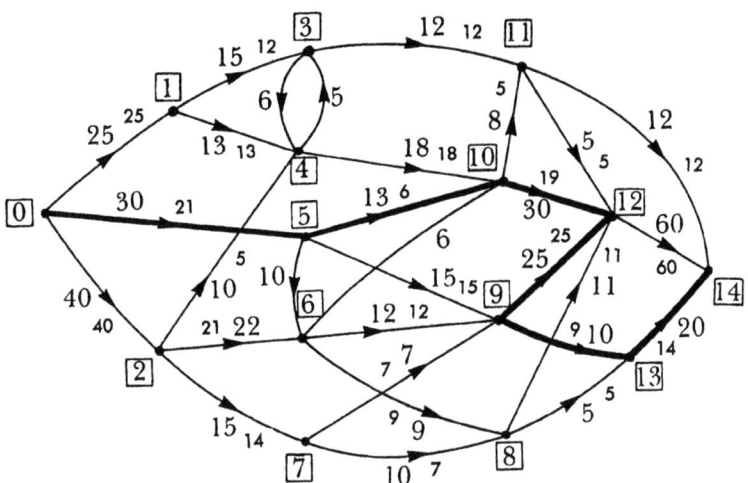

TABLE VII.5.12

j	1	2	3	4	5	6	7	8	9	10	11	12	13	14
t_j	13	21	5	5	9	10	1	3	25	7	3	11	1	6
k_j	4	6	4	2	0	5	2	7	12	5	10	10	9	13

FIGURE VII.5.18. Solution (maximal flow) for Example VII.5.7. W = 88.

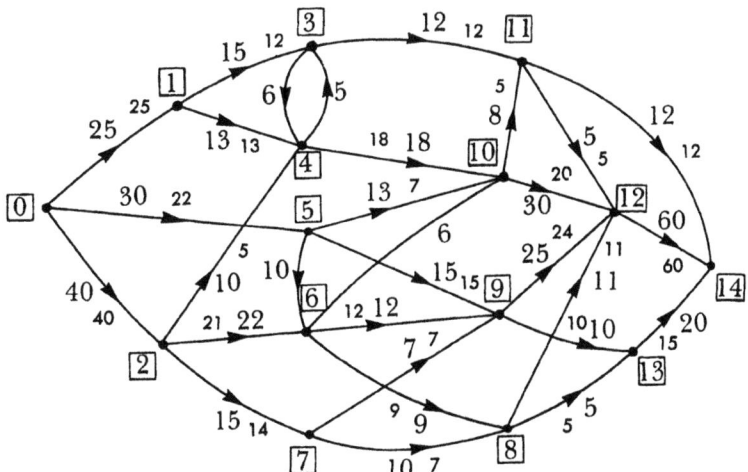

This last figure (VII.5.18) gives us the solution: the sink cannot be labeled. This may be checked directly by carrying out the labeling procedure. A shorter way of checking is to note that the flow in the network is now equal to 87 units. Consider then the cut

$$L = \{0, 1, 2, ..., 12\} \qquad U = \{13,14\}$$

There are 4 arcs leading from L into U: these are (8,13), (9,13), (11,14) and (12,14), with capacities 5, 10, 12, and 60 respectively. The value of the cut (the sum of these capacities) is 87, and so $w = v(L, U)$. By the max-flow min-cut theorem, we conclude that this is the maximum possible flow.

We have, throughout this chapter, relied considerably on the graphic technique. This is a good idea, inasmuch as the graph of a network is readily understood and helps to give a good idea of the situation. On the other hand, graphs are not necessary. It is generally possible to replace a graph by a matrix, and, for purposes of storing data (using a computer, which perhaps cannot appreciate diagrams) this may actually be necessary. The network of Example VII.5.7 can be represented by the following matrix, which gives the capacities of the several arcs and also serves to give the excess capacities when there is no flow in the network.

	1	2	3	4	5	6	7	8	9	10	11	12	13	14
0	25	40	0	0	30	0	0	0	0	0	0	0	0	0
1	0	0	15	13	0	0	0	0	0	0	0	0	0	0
2	0	0	0	10	0	22	15	0	0	0	0	0	0	0
3	0	0	0	6	0	0	0	0	0	0	12	0	0	0
4	0	0	5	0	0	0	0	0	0	18	0	0	0	0
5	0	0	0	0	0	10	0	0	15	13	0	0	0	0
6	0	0	0	0	0	0	0	9	12	6	0	0	0	0
7	0	0	0	0	0	0	0	10	7	0	0	0	0	0
8	0	0	0	0	0	0	0	0	0	0	0	11	5	0
9	0	0	0	0	0	0	0	0	0	0	0	25	10	0
10	0	0	0	0	0	0	0	0	0	0	8	30	0	0
11	0	0	0	0	0	0	0	0	0	0	0	5	0	12
12	0	0	0	0	0	0	0	0	0	0	0	0	0	60
13	0	0	0	0	0	0	0	0	0	0	0	0	0	20

It is possible, from this matrix, to carry out the labeling procedure without needing to look at a graph. Two additional rows of the matrix would be sufficient for this purpose. After the additional flow has been decided, a new matrix of excess capacities can be formed, according to the relation

$$g'_{ij} = g_{ij} - \Delta_{ij} + \Delta_{ji}$$

where g'_{ij} and g_{ij} are the new and old excess capacities, respectively, while Δ_{ij} and Δ_{ji} are the additional flows from i to j and from j to i, respectively. Corresponding to the network of Figure VII.5.6, we have the matrix

	1	2	3	4	5	6	7	8	9	10	11	12	13	14
0	13	40	0	0	30	0	0	0	0	0	0	0	0	0
1	0	0	3	13	0	0	0	0	0	0	0	0	0	0
2	0	0	0	10	0	22	15	0	0	0	0	0	0	0
3	12	0	0	6	0	0	0	0	0	0	0	0	0	0
4	0	0	5	0	0	0	0	0	0	18	0	0	0	0
5	0	0	0	0	0	10	0	0	15	13	0	0	0	0
6	0	0	0	0	0	0	0	9	12	6	0	0	0	0
7	0	0	0	0	0	0	0	10	7	0	0	0	0	0
8	0	0	0	0	0	0	0	0	0	0	0	11	5	0
9	0	0	0	0	0	0	0	0	0	0	0	25	10	0
10	0	0	0	0	0	0	0	0	0	0	8	30	0	0
11	0	0	12	0	0	0	0	0	0	0	0	5	0	0
12	0	0	0	0	0	0	0	0	0	0	0	0	0	60
13	0	0	0	0	0	0	0	0	0	0	0	0	0	20

The procedure can continue, using matrices only, in this manner. There is no difference, conceptually, between the matric and the graphic procedures.

The reader is invited to construct the matrices of excess capacities corresponding to Figures VII.5.7 through VII.5.18.

PROBLEMS ON MAXIMAL FLOWS THROUGH NETWORKS

1. Find the maximal flows in the networks in figures VII.5.19 to VII.5.23. What are the minimal cuts in each case?

FIGURE VII.5.19.

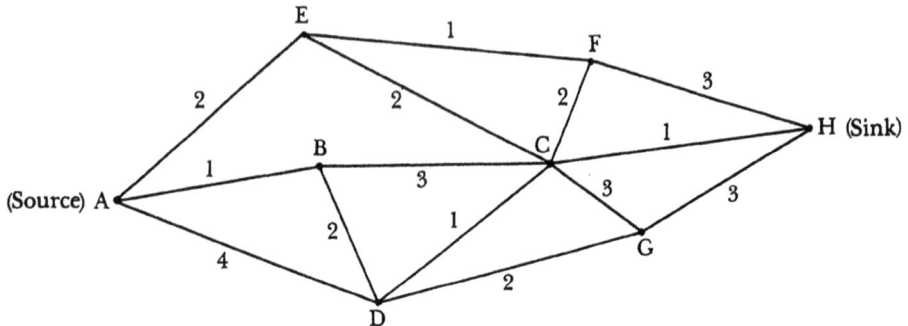

FIGURE VII.5.20..

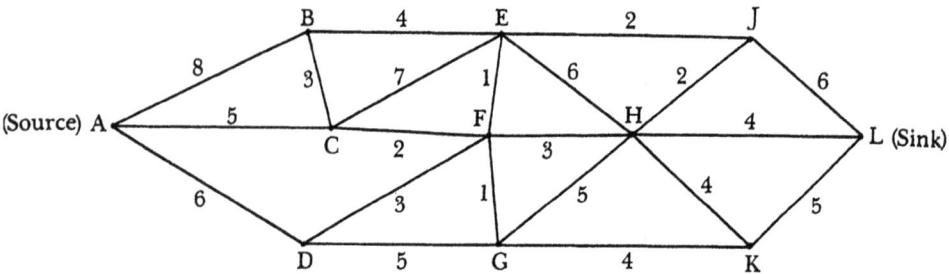

FIGURE VII.5.21.

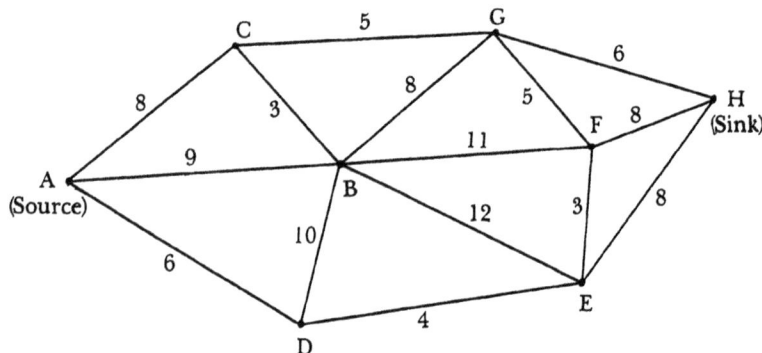

FIGURE VII.5.22. Flow is allowed only in the direction of the arrows.

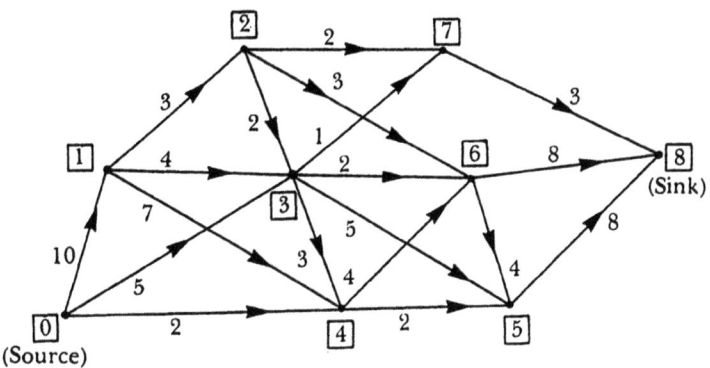

FIGURE VII.5.23. Flow is allowed only in the direction of the arrows.

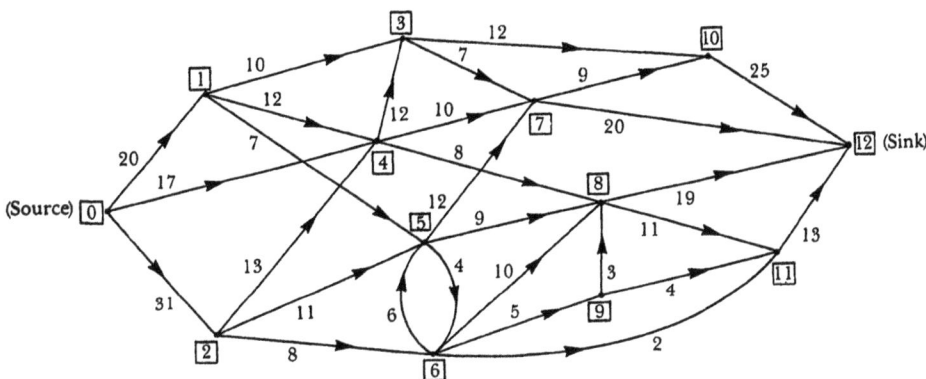

2. A steel company wishes to transport ore from the mine to the refinery. It can do this by shipping from the mine to any one of several intermediate cities, and on to the plant. The available space (in tons of ore) on the carriers between cities is given by the following table (where city 0 is the mine, and 13 is the refinery).

	0	1	2	3	4	5	6	7	8	9	10	11	12	13
0	-	30	40	25	0	0	0	0	0	0	0	0	0	0
1	10	-	0	0	12	17	0	0	0	0	0	0	0	0
2	8	0	-	0	22	15	8	0	0	0	0	0	0	0
3	12	0	0	-	17	12	0	0	0	0	0	0	0	0
4	0	0	2	0	-	0	0	15	27	0	0	0	0	0
5	0	3	0	5	0	-	15	12	0	10	8	0	0	0
6	0	0	4	0	0	0	-	16	0	4	10	0	0	0
7	0	0	0	0	0	8	0	-	12	8	0	0	0	2
8	0	0	0	0	6	0	0	0	-	0	0	25	0	0
9	0	0	0	0	0	1	10	5	0	-	0	6	14	0
10	0	0	0	0	0	0	3	0	0	0	-	0	15	0
11	0	0	0	0	0	0	0	0	0	8	0	-	0	30
12	0	0	0	0	0	0	0	0	0	0	6	0	-	40
13	0	0	0	0	0	0	0	0	0	0	0	5	8	-

What is the maximal amount of ore than can be transported?

GPSR Compliance
The European Union's (EU) General Product Safety Regulation (GPSR) is a set of rules that requires consumer products to be safe and our obligations to ensure this.

If you have any concerns about our products, you can contact us on

ProductSafety@springernature.com

In case Publisher is established outside the EU, the EU authorized representative is:

Springer Nature Customer Service Center GmbH
Europaplatz 3
69115 Heidelberg, Germany

www.ingramcontent.com/pod-product-compliance
Ingram Content Group UK Ltd.
Pitfield, Milton Keynes, MK11 3LW, UK
UKHW022230230426
12048UKWH00016BA/1167